Probability and Statistics for
Engineering and the Sciences

EIGHTH EDITION

Jay L. Devore

Prepared by

Matthew A. Carlton
Cal Poly State University

BROOKS/COLE
CENGAGE Learning

Australia • Brazil • Japan • Korea • Mexico • Singapore • Spain • United Kingdom • United States

For product information and technology assistance, contact us at **Cengage Learning Customer & Sales Support, 1-800-354-9706**

For permission to use material from this text or product, submit all requests online at **www.cengage.com/permissions** Further permissions questions can be emailed to **permissionrequest@cengage.com**

ISBN-13: 978-0-8400-6539-1
ISBN-10: 0-8400-6539-6

Brooks/Cole
20 Channel Center Street
Boston, MA 02210
USA

Cengage Learning is a leading provider of customized learning solutions with office locations around the globe, including Singapore, the United Kingdom, Australia, Mexico, Brazil, and Japan. Locate your local office at: **www.cengage.com/global**

Cengage Learning products are represented in Canada by Nelson Education, Ltd.

To learn more about Brooks/Cole, visit **www.cengage.com/brookscole**

Purchase any of our products at your local college store or at our preferred online store **www.cengagebrain.com**

Printed in the United States of America
1 2 3 4 5 6 7 15 14 13 12 11

CONTENTS

CHAPTER 1

Section 1.1

1.

    a. *Los Angeles Times, Oberlin Tribune, Gainesville Sun, Washington Post*

    b. Duke Energy, Clorox, Seagate, Neiman Marcus

    c. Vince Correa, Catherine Miller, Michael Cutler, Ken Lee

    d. 2.97, 3.56, 2.20, 2.97

3.

    a. How likely is it that more than half of the sampled computers will need or have needed warranty service? What is the expected number among the 100 that need warranty service? How likely is it that the number needing warranty service will exceed the expected number by more than 10?

    b. Suppose that 15 of the 100 sampled needed warranty service. How confident can we be that the proportion of *all* such computers needing warranty service is between .08 and .22? Does the sample provide compelling evidence for concluding that more than 10% of all such computers need warranty service?

5.

    a. No. All students taking a large statistics course who participate in an SI program of this sort.

    b. The advantage to randomly allocating students to the two groups is that the two groups should then be fairly comparable before the study. If the two groups perform differently in the class, we might attribute this to the treatments (SI and control). If it were left to students to choose, stronger or more dedicated students might gravitate toward SI, confounding the results.

    c. If all students were put in the treatment group, there would be no firm basis for assessing the effectiveness of SI (nothing to which the SI scores could reasonably be compared).

7.   One could generate a simple random sample of all single-family homes in the city, or a stratified random sample by taking a simple random sample from each of the 10 district neighborhoods. From each of the selected homes, values of all desired variables would be determined. This would be an enumerative study because there exists a finite, identifiable population of objects from which to sample.

1

9.

    a. There could be several explanations for the variability of the measurements. Among them could be measurement error (due to mechanical or technical changes across measurements), recording error, differences in weather conditions at time of measurements, etc.

    b. No, because there is no sampling frame.

Section 1.2

11.

6L	034
6H	667899
7L	00122244
7H	
8L	001111122344
8H	5557899
9L	03
9H	58

stem: tens
leaf : ones

This display brings out the gap in the data—there are no scores in the high 70's.

13.

    a.

12	2
12	445
12	6667777
12	889999
13	00011111111
13	222222222233333333333333333
13	4444444444444444445555555555555555555
13	6666666666667777777777
13	888888888888999999
14	0000001111
14	2333333
14	444
14	77

stem: tens
leaf: ones

The observations are highly concentrated at around 134 or 135, where the display suggests the typical value falls.

2

b.

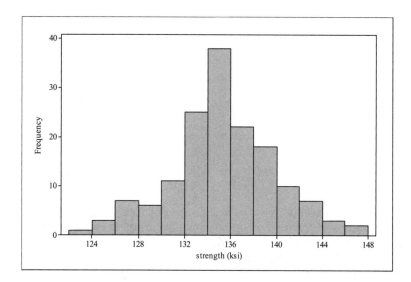

The histogram of ultimate strengths is symmetric and unimodal, with the point of symmetry at approximately 135 ksi. There is a moderate amount of variation, and there are no gaps or outliers in the distribution.

15.

American		French
	8	1
755543211000	9	00234566
9432	10	2356
6630	11	1369
850	12	223558
8	13	7
	14	
	15	8
2	16	

American movie times are unimodal strongly positively skewed, while French movie times appear to be bimodal. A typical American movie runs about 95 minutes, while French movies are typically either around 95 minutes or around 125 minutes. American movies are generally shorter than French movies and are less variable in length. Finally, both American and French movies occasionally run very long (outliers at 162 minutes and 158 minutes, respectively, in the samples).

17.

a.

Number	Nonconforming Frequency	Relative Frequency (Freq./60)
0	7	0.117
1	12	0.200
2	13	0.217
3	14	0.233
4	6	0.100
5	3	0.050
6	3	0.050
7	1	0.017
8	1	0.017
		1.001 *rounding error*

b. The number of batches with at most 5 nonconforming items is 7+12+13+14+6+3 = 55, which is a proportion of 55/60 = .917. The proportion of batches with (strictly) fewer than 5 nonconforming items is 52/60 = .867. Notice that these proportions could also have been computed by using the relative frequencies: e.g., proportion of batches with 5 or fewer nonconforming items = 1– (.05+.017+.017) = .916; proportion of batches with fewer than 5 nonconforming items = 1 – (.05+.05+.017+.017) = .866.

c. The center of the histogram is somewhere around 2 or 3, and there is some positive skewness in the data. The histogram also shows that there is a lot of spread/variation in this data.

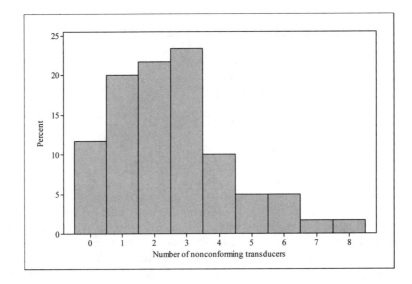

19.

    a. From this frequency distribution, the proportion of wafers that contained at least one particle is (100-1)/100 = .99, or 99%. Note that it is much easier to subtract 1 (which is the number of wafers that contain 0 particles) from 100 than it would be to add all the frequencies for 1, 2, 3,… particles. In a similar fashion, the proportion containing at least 5 particles is (100 - 1-2-3-12-11)/100 = 71/100 = .71, or, 71%.

    b. The proportion containing between 5 and 10 particles is (15+18+10+12+4+5)/100 = 64/100 = .64, or 64%. The proportion that contain strictly between 5 and 10 (meaning strictly *more* than 5 and strictly *less* than 10) is (18+10+12+4)/100 = 44/100 = .44, or 44%.

    c. The following histogram was constructed using Minitab. The histogram is *almost* symmetric and unimodal; however, the distribution has a few smaller modes and has a very slight positive skew.

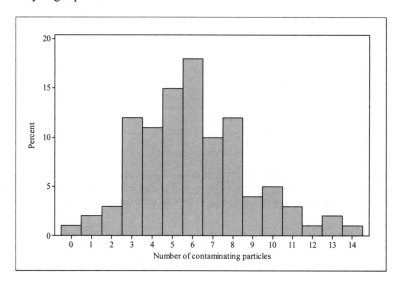

21.

    a. A histogram of the y data appears below. From this histogram, the number of subdivisions having no cul-de-sacs (i.e., $y = 0$) is 17/47 = .362, or 36.2%. The proportion having at least one cul-de-sac ($y \geq 1$) is (47 − 17)/47 = 30/47 = .638, or 63.8%. Note that subtracting the number of cul-de-sacs with $y = 0$ from the total, 47, is an easy way to find the number of subdivisions with $y \geq 1$.

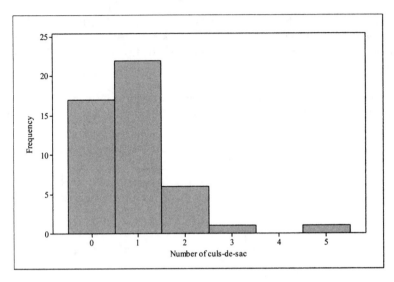

    b. A histogram of the z data appears below. From this histogram, the number of subdivisions with at most 5 intersections (i.e., $z \leq 5$) is 42/47 = .894, or 89.4%. The proportion having fewer than 5 intersections (i.e., $z < 5$) is 39/47 = .830, or 83.0%.

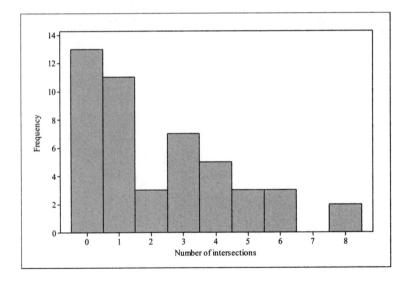

23.   Note: since the class intervals have unequal length, we must use a *density scale*.

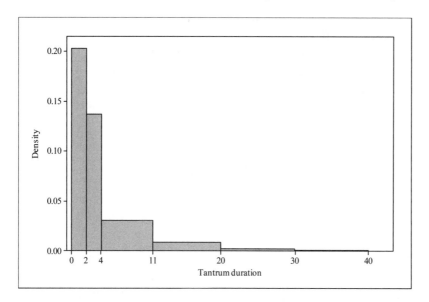

The distribution of tantrum durations is unimodal and heavily positively skewed. Most tantrums last between 0 and 11 minutes, but a few last more than half an hour! With such heavy skewness, it's difficult to give a representative value.

25.   The transformation creates a much more symmetric, mound-shaped histogram.

Histogram of original data:

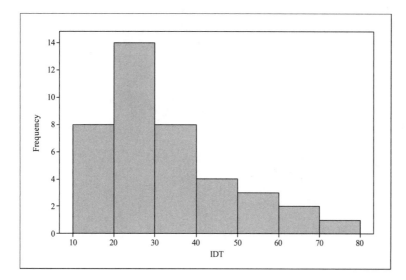

7

Histogram of transformed data:

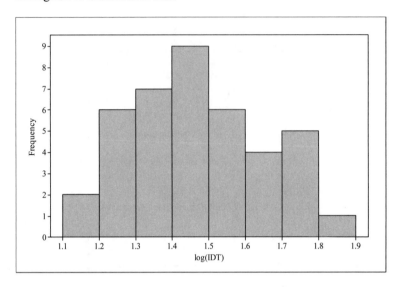

27.

    a. The endpoints of the class intervals overlap. For example, the value 50 falls in both of the intervals 0–50 and 50–100.

    b. The lifetime distribution is positively skewed. A representative value is around 100. There is a great deal of variability in lifetimes and several possible candidates for outliers.

Class Interval	Frequency	Relative Frequency
0–< 50	9	0.18
50–<100	19	0.38
100–<150	11	0.22
150–<200	4	0.08
200–<250	2	0.04
250–<300	2	0.04
300–<350	1	0.02
350–<400	1	0.02
400–<450	0	0.00
450–<500	0	0.00
500–<550	1	0.02
	50	1.00

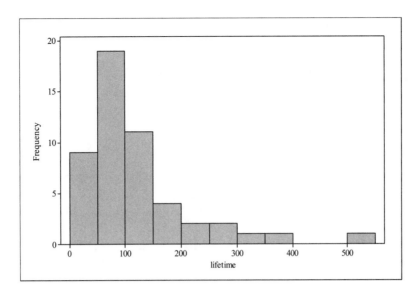

c. There is much more symmetry in the distribution of the transformed values than in the values themselves, and less variability. There are no longer gaps or obvious outliers.

Class Interval	Frequency	Relative Frequency
2.25–<2.75	2	0.04
2.75–<3.25	2	0.04
3.25–<3.75	3	0.06
3.75–<4.25	8	0.16
4.25–<4.75	18	0.36
4.75–<5.25	10	0.20
5.25–<5.75	4	0.08
5.75–<6.25	3	0.06

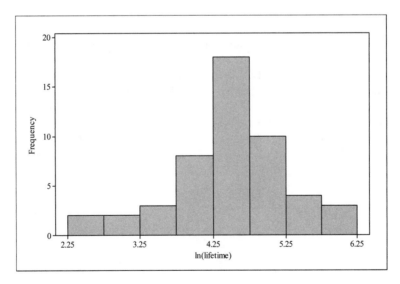

d. The proportion of lifetime observations in this sample that are less than 100 is .18 + .38 = .56, and the proportion that is at least 200 is .04 + .04 + .02 + .02 + .02 = .14.

9

29.

Complaint	Frequency	Relative Frequency
B	7	0.1167
C	3	0.0500
F	9	0.1500
J	10	0.1667
M	4	0.0667
N	6	0.1000
O	21	0.3500
	60	1.0000

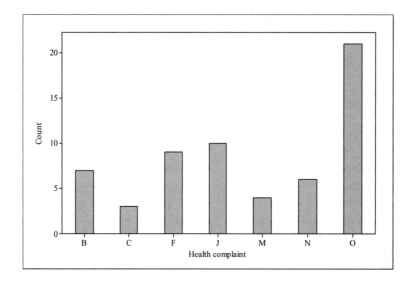

31.

Class	Frequency	Cum. Freq.	Cum. Rel. Freq.
0.0–<4.0	2	2	0.050
4.0–<8.0	14	16	0.400
8.0–<12.0	11	27	0.675
12.0–<16.0	8	35	0.875
16.0–<20.0	4	39	0.975
20.0–<24.0	0	39	0.975
24.0–<28.0	1	40	1.000

Section 1.3

33.

    a. Using software, $\bar{x} = 640.5$ ($\$640,500$) and $\tilde{x} = 582.5$ ($\$582,500$). The average sale price for a home in this sample was $\$640,500$. Half the sales were for less than $\$582,500$, while half were for more than $\$582,500$.

    b. Changing that one value lowers the sample mean to 610.5 ($\$610,500$) but has no effect on the sample median.

    c. After removing the two largest and two smallest values, $\bar{x}_{tr(20)} = 591.2$ ($\$591,200$).

    d. A 10% trimmed mean from removing just the highest and lowest values is $\bar{x}_{tr(10)} = 596.3$. To form a 15% trimmed mean, take the average of the 10% and 20% trimmed means to get $\bar{x}_{tr(15)} = (591.2 + 596.3)/2 = 593.75$ ($\$593,750$).

35.

    a. From software, the sample mean is $\bar{x} = 12.55$ and the sample median is $\tilde{x} = 12.5$. The 12.5% trimmed mean requires that we first trim $(.125)(n) = 1$ value from the ends of the ordered data set. Then we average the remaining 6 values. The 12.5% trimmed mean $\bar{x}_{tr(12.5)}$ is $74.4/6 = 12.4$.

    All three measures of center are similar, indicating little skewness to the data set.

    b. The smallest value, 8.0, could be increased to any number below 12.0 (a change of less than 4.0) without affecting the value of the sample median.

    c. The values obtained in part (a) can be used directly. For example, the sample mean of 12.55 psi could be re-expressed as

$$12.55 \text{ psi} \times \left(\frac{1 \text{ ksi}}{2.2 \text{ psi}} \right) = 5.705 \text{ ksi}$$

37. $\bar{x} = 12.01$, $\tilde{x} = 11.35$, $\bar{x}_{tr(10)} = 11.46$. The median or the trimmed mean would be better choices than the mean because of the outlier 21.9.

39.

    a. $\Sigma x_i = 16.475$ so $\bar{x} = \dfrac{16.475}{16} = 1.0297$; $\tilde{x} = \dfrac{(1.007 + 1.011)}{2} = 1.009$

    b. 1.394 can be decreased until it reaches 1.011 (i.e. by $1.394 - 1.011 = 0.383$), the largest of the 2 middle values. If it is decreased by more than 0.383, the median will change.

11

41.

    a. $x/n = 7/10 = .7$

    b. $\bar{x} = .70 =$ the sample proportion of successes

    c. To have x/n equal .80 requires $x/25 = .80$ or $x = (.80)(25) = 20$. There are 7 successes (S) already, so another $20 - 7 = 13$ would be required.

43. The median and certain trimmed means can be calculated, while the mean cannot — the exact values of the "100+" observations are required to calculate the mean. $\tilde{x} = \dfrac{(57 + 79)}{2} = 68.0$, $\bar{x}_{tr(20)} = 66.2$, $\bar{x}_{tr(30)} = 67.5$.

Section 1.4

45.

    a. $\bar{x} = 115.58$. The deviations from the mean are $116.4 - 115.58 = .82$, $115.9 - 115.58 = .32$, $114.6 - 115.58 = -.98$, $115.2 - 115.58 = -.38$, and $115.8 - 115.58 = .22$. Notice that the deviations from the mean sum to zero, as they should.

    b. $s^2 = [(.82)^2 + (.32)^2 + (-.98)^2 + (-.38)^2 + (.22)^2]/(5 - 1) = 1.928/4 = .482$, so $s = .694$.

    c. $\Sigma x_i^2 = 66795.61$, so $s^2 = S_{xx}/(n - 1) = \left(\Sigma x_i^2 - (\Sigma x_i)^2 / n\right)/(n - 1) = (66795.61 - (577.9)^2/5)/4 = 1.928/4 = .482$.

    d. The new sample values are: 16.4 15.9 14.6 15.2 15.8. While the new mean is 15.58, all the deviations are the same as in part (a), and the variance of the transformed data is identical to that of part (b).

47. From software, $\tilde{x} = 109.5$ MPa, $\bar{x} = 116.2$ MPa, and $s = 25.75$ MPa. Half the fracture strength measurements in the sample are below 109.5 MPa, and half are above. On average, we would expect a fracture strength of 116.2 MPa. In general, the size of a typical deviation from the sample mean (116.2) is about 25.75 MPa. Some observations may deviate from 116.2 by more than this and some by less.

49.

    a. $\Sigma x_i = 2.75 + \cdots + 3.01 = 56.80$, $\Sigma x_i^2 = 2.75^2 + \cdots + 3.01^2 = 197.8040$

    b. $s^2 = \dfrac{197.8040 - (56.80)^2/17}{16} = \dfrac{8.0252}{16} = .5016$, $s = .708$

12

51.

    a. From software, $s^2 = 1264.77$ min^2 and $s = 35.56$ min. Working by hand, $\Sigma x = 2563$ and $\Sigma x^2 = 368501$, so

$$s^2 = \frac{368501 - (2563)^2 / 19}{19 - 1} = 1264.766 \text{ and } s = \sqrt{1264.766} = 35.564$$

    b. If y = time in hours, then $y = cx$ where $c = \frac{1}{60}$. So, $s_y^2 = c^2 s_x^2 = \left(\frac{1}{60}\right)^2 1264.766 = .351$ hr^2 and $s_y = cs_x = \left(\frac{1}{60}\right) 35.564 = .593$ hr.

53.

    a. Using software, for the sample of balanced funds we have $\bar{x} = 1.121, \tilde{x} = 1.050, s = 0.536$; for the sample of growth funds we have $\bar{x} = 1.244, \tilde{x} = 1.100, s = 0.448$.

    b. The distribution of expense ratios for this sample of balanced funds is fairly symmetric, while the distribution for growth funds is positively skewed. These balanced and growth mutual funds have similar median expense ratios (1.05% and 1.10%, respectively), but expense ratios are generally higher for growth funds. The lone exception is a balanced fund with a 2.86% expense ratio. (There is also one unusually low expense ratio in the sample of balanced funds, at 0.09%.)

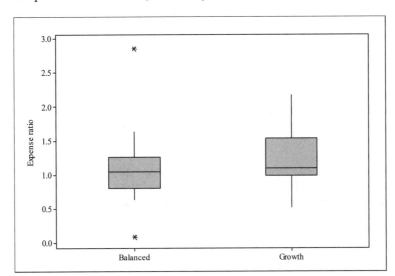

55.

    a. Lower half of the data set: 325 325 334 339 356 356 359 359 363 364 364 366 369, whose median, and therefore the lower fourth, is 359 (the 7th observation in the sorted list).

       Upper half of the data set: 370 373 373 374 375 389 392 393 394 397 402 403 424, whose median, and therefore the upper fourth is 392.

       So, $f_s = 392 - 359 = 33$.

13

b. inner fences: $359 - 1.5(33) = 309.5$, $392 + 1.5(33) = 441.5$
To be a mild outlier, an observation must be below 309.5 or above 441.5. There are none in this data set. Clearly, then, there are also no extreme outliers.

c. A boxplot of this data appears below. The distribution of escape times is roughly symmetric with no outliers. Notice the box plot "hides" the fact that the distribution contains two gaps, which can be seen in the stem-and-leaf display.

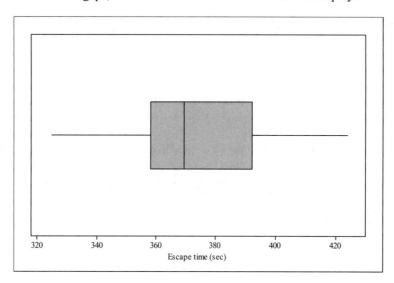

Escape time (sec)

d. Not until the value $x = 424$ is lowered below the upper fourth value of 392 would there be any change in the value of the upper fourth (and, thus, of the fourth spread). That is, the value $x = 424$ could not be decreased by more than $424 - 392 = 32$ seconds.

57.

a. $f_s = 216.8 - 196.0 = 20.8$
inner fences: $196 - 1.5(20.8) = 164.6$, $216.8 + 1.5(20.8) = 248$
outer fences: $196 - 3(20.8) = 133.6$, $216.8 + 3(20.8) = 279.2$
Of the observations listed, 125.8 is an extreme low outlier and 250.2 is a mild high outlier.

b. A boxplot of this data appears below. There is a bit of positive skew to the data but, except for the two outliers identified in part (a), the variation in the data is relatively small.

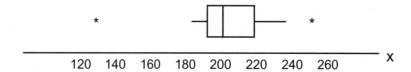

14

59.

    a. If you aren't using software, don't forget to *sort* the data first!
ED: median = .4, lower fourth = (.1 + .1)/2 = .1, upper fourth = (2.7 + 2.8)/2 = 2.75, fourth spread = 2.75 − .1 = 2.65

    Non-ED: median = (1.5 + 1.7)/2 = 1.6, lower fourth = .3, upper fourth = 7.9, fourth spread = 7.9 − .3 = 7.6.

    b. *ED*: mild outliers are less than .1 − 1.5(2.65) = −3.875 or greater than 2.75 + 1.5(2.65) = 6.725. Extreme outliers are less than .1 − 3(2.65) = −7.85 or greater than 2.75 + 3(2.65) = 10.7. So, the two largest observations (11.7, 21.0) are extreme outliers and the next two largest values (8.9, 9.2) are mild outliers. There are no outliers at the lower end of the data.

    Non-ED: mild outliers are less than .3 − 1.5(7.6) = −11.1 or greater than 7.9 + 1.5(7.6) = 19.3. Note that there are no mild outliers in the data, hence there cannot be any extreme outliers, either.

    c. A comparative boxplot appears below. The outliers in the ED data are clearly visible. There is noticeable positive skewness in both samples; the Non-ED sample has more variability then the Ed sample; the typical values of the ED sample tend to be smaller than those for the Non-ED sample.

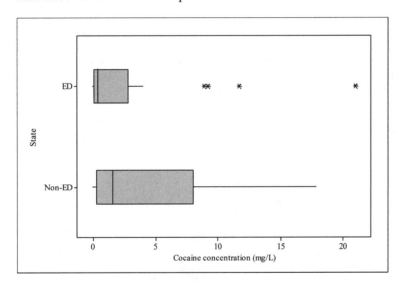

61. Outliers occur in the 6a.m. data. The distributions at the other times are fairly symmetric. Variability and the "typical" gasoline-vapor coefficient values increase somewhat until 2p.m., then decrease slightly.

Supplementary Exercises

63.   As seen in the histogram below, this noise distribution is bimodal (but close to unimodal) with a positive skew and no outliers. The mean noise level is 64.89 dB and the median noise level is 64.7 dB. The fourth spread of the noise measurements is about $70.4 - 57.8 = 12.6$ dB.

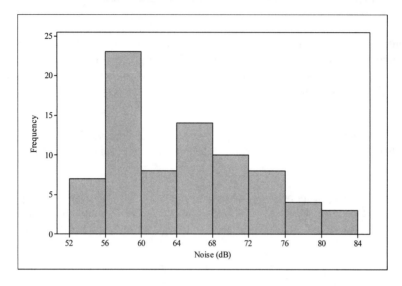

65.

   a.   The histogram appears below.   A representative value for this data would be around 90 MPa. The histogram is reasonably symmetric, unimodal, and somewhat bell-shaped with a fair amount of variability ($s \approx 3$ or 4 MPa).

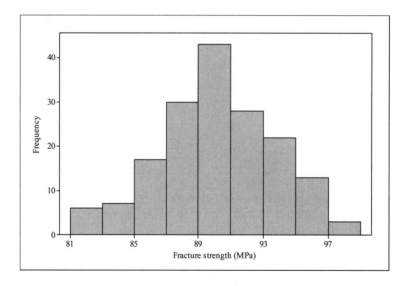

   b.   The proportion of the observations that are at least 85 is $1 - (6+7)/169 = .9231$. The proportion less than 95 is $1 - (13+3)/169 = .9053$.

16

Chapter 1: Overview and Descriptive Statistics

c. 90 is the midpoint of the class 89–<91, which contains 43 observations (a relative frequency of 43/169 = .2544). Therefore about half of this frequency, .1272, should be added to the relative frequencies for the classes to the left of $x = 90$. That is, the approximate proportion of observations that are less than 90 is .0355 + .0414 + .1006 + .1775 + .1272 = .4822.

67.

a. Aortic root diameters for males have mean 3.64 cm, median 3.70 cm, standard deviation 0.269 cm, and fourth spread 0.40. The corresponding values for females are $\bar{x} = 3.28$ cm, $\tilde{x} = 3.15$ cm, $s = 0.478$ cm, and $f_s = 0.50$ cm. Aortic root diameters are typically (though not universally) somewhat smaller for females than for males, and females show more variability. The distribution for males is negatively skewed, while the distribution for females is positively skewed (see graphs below).

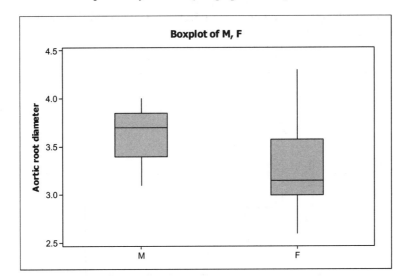

b. For females ($n = 10$), the 10% trimmed mean is the average of the middle 8 observations: $\bar{x}_{tr(10)} = 3.24$ cm. For males ($n = 13$), the 1/13 trimmed mean is 40.2/11 = 3.6545, and the 2/13 trimmed mean is 32.8/9 = 3.6444. Interpolating, the 10% trimmed mean is $\bar{x}_{tr(10)} = 0.7(3.6545) + 0.3(3.6444) = 3.65$ cm. (10% is three-tenths of the way from 1/13 to 2/13).

69.

a.

$$\bar{y} = \frac{\sum y_i}{n} = \frac{\sum(ax_i + b)}{n} = \frac{a\sum x_i + \sum b}{n} = \frac{a\sum x_i + nb}{n} = a\bar{x} + b$$

$$s_y^2 = \frac{\sum(y_i - \bar{y})^2}{n-1} = \frac{\sum(ax_i + b - (a\bar{x}+b))^2}{n-1} = \frac{\sum(ax_i - a\bar{x})^2}{n-1}$$

$$= \frac{a^2\sum(x_i - \bar{x})^2}{n-1} = a^2 s_x^2$$

17

b.

$$x = {}^{\circ}C, y = {}^{\circ}F$$

$$\bar{y} = \frac{9}{5}(87.3) + 32 = 189.14\,{}^{\circ}F$$

$$s_y = \sqrt{s_y^2} = \sqrt{\left(\frac{9}{5}\right)^2 (1.04)^2} = \sqrt{3.5044} = 1.872\,{}^{\circ}F$$

71.

    a. The mean, median, and trimmed mean are virtually identical, which suggests symmetry. If there are outliers, they are balanced. The range of values is only 25.5, but half of the values are between 132.95 and 138.25.

    b. See the comments for (a). In addition, using 1.5(Q3 – Q1) as a yardstick, the two largest and three smallest observations are mild outliers.

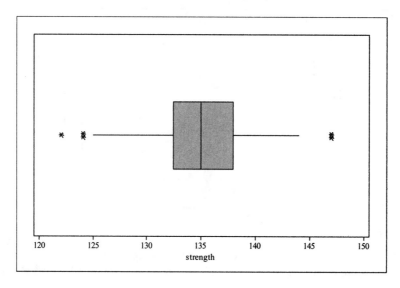

73. From software, $\bar{x} = .9255$, $s = .0809$; $\tilde{x} = .93$, $f_s = .1$. The cadence observations are slightly skewed (mean = .9255 strides/sec, median = .93 strides/sec) and show a small amount of variability (standard deviation = .0809, fourth spread = .1). There are no apparent outliers in the data.

```
7 8                  stem = tenths
8 11556              leaf = hundredths
9 2233335566
0 0566
```

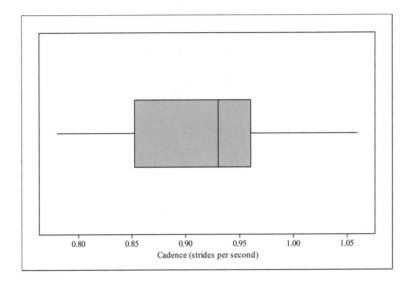

75.

    a. The median is the same (371) in each plot and all three data sets are very symmetric. In addition, all three have the same minimum value (350) and same maximum value (392). Moreover, all three data sets have the same lower (364) and upper quartiles (378). So, all three boxplots will be *identical*. (Slight differences in the boxplots below are due to the way Minitab software interpolates to calculate the quartiles.)

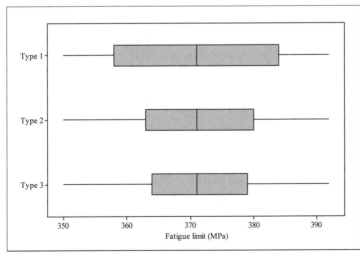

19

b. A comparative dotplot is shown below. These graphs show that there are differences in the variability of the three data sets. They also show differences in the way the values are distributed in the three data sets, especially big differences in the presence of gaps and clusters.

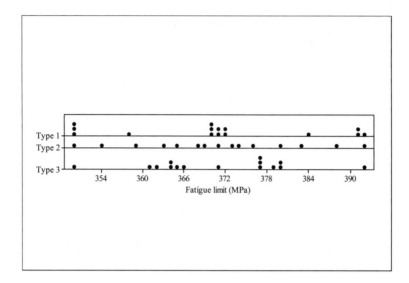

c. The boxplot in (a) is not capable of detecting the differences among the data sets. The primary reason is that boxplots give up some detail in describing data because they use only five summary numbers for comparing data sets.

77.

   a.

```
0    444444444577888999          leaf = 1.0
1    00011111111124455669999     stem = 0.1
2    1234457
3    11355
4    17
5    3
6
7    67
8    1

HI   10.44, 13.41
```

b. Since the intervals have unequal width, you must use a *density scale.*

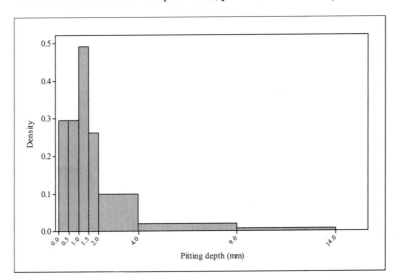

c. Representative depths are quite similar for the three types of soils — between 1.5 and 2. Data from the C and CL soils shows much more variability than for the other two types. The boxplots for the first three types show substantial positive skewness both in the middle 50% and overall. The boxplot for the SYCL soil shows negative skewness in the middle 50% and mild positive skewness overall. Finally, there are multiple outliers for the first three types of soils, including extreme outliers.

79.

a. $\displaystyle\sum_{i=1}^{n+1} x_i = \sum_{i=1}^{n} x_i + x_{n+1} = n\overline{x}_n + x_{n+1}$, so $\displaystyle\overline{x}_{n+1} = \frac{1}{n+1}\sum_{i=1}^{n+1} x_i = \frac{n\overline{x}_n + x_{n+1}}{n+1}$.

b. In the second line below, we artificially add and subtract $n\overline{x}_n^2$ to create the term needed for the sample variance:

$$ns_{n+1}^2 = \sum_{i=1}^{n+1}(x_i - \overline{x}_{n+1})^2 = \sum_{i=1}^{n+1} x_i^2 - (n+1)\overline{x}_{n+1}^2$$

$$= \sum_{i=1}^{n} x_i^2 + x_{n+1}^2 - (n+1)\overline{x}_{n+1}^2 = \left[\sum_{i=1}^{n} x_i^2 - n\overline{x}_n^2\right] + n\overline{x}_n^2 + x_{n+1}^2 - (n+1)\overline{x}_{n+1}^2$$

$$= (n-1)s_n^2 + \left\{x_{n+1}^2 + n\overline{x}_n^2 - (n+1)\overline{x}_{n+1}^2\right\}$$

Substitute the expression for $\overline{x}_{n+1}$ from part (a) into the expression in braces, and it simplifies to $\dfrac{n}{n+1}(x_{n+1} - \overline{x}_n)^2$, as desired.

c. First, $\overline{x}_{16} = \dfrac{15(12.58)+11.8}{16} = \dfrac{200.5}{16} = 12.53$. Then, solving (b) for s_{n+1}^2 gives

$s_{n+1}^2 = \dfrac{n-1}{n}s_n^2 + \dfrac{1}{n+1}(x_{n+1} - \overline{x}_n)^2 = \dfrac{14}{15}(.512)^2 + \dfrac{1}{16}(11.8-12.58)^2 = .238$. Finally, the

standard deviation is $s_{16} = \sqrt{.238} = .532$.

21

81. Assuming that the histogram is unimodal, then there is evidence of positive skewness in the data since the median lies to the left of the mean (for a symmetric distribution, the mean and median would coincide).

For more evidence of skewness, compare the distances of the 5th and 95th percentiles from the median: median – 5th %ile = 500 – 400 = 100, while 95th %ile – median = 720 – 500 = 220. Thus, the largest 5% of the values (above the 95th percentile) are further from the median than are the lowest 5%. The same skewness is evident when comparing the 10th and 90th percentiles to the median, or comparing the maximum and minimum to the median.

83.

a. When there is perfect symmetry, the smallest observation y_1 and the largest observation y_n will be equidistant from the median, so $y_n - \tilde{x} = \tilde{x} - y_1$. Similarly, the second-smallest and second-largest will be equidistant from the median, so $y_{n-1} - \tilde{x} = \tilde{x} - y_2$, and so on. Thus, the first and second numbers in each pair will be equal, so that each point in the plot will fall exactly on the 45° line.

When the data is positively skewed, y_n will be much further from the median than is y_1, so $y_n - \tilde{x}$ will considerably exceed $\tilde{x} - y_1$ and the point $(y_n - \tilde{x}, \tilde{x} - y_1)$ will fall considerably below the 45° line, as will the other points in the plot.

b. The median of these $n = 26$ observations is 221.6 (the midpoint of the 13th and 14th ordered values). The first point in the plot is (2745.6 – 221.6, 221.6 – 4.1) = (2524.0, 217.5). The others are: (1476.2, 213.9), (1434.4, 204.1), (756.4, 190.2), (481.8, 188.9), (267.5, 181.0), (208.4, 129.2), (112.5, 106.3), (81.2, 103.3), (53.1, 102.6), (53.1, 92.0), (33.4, 23.0), and (20.9, 20.9). The first number in each of the first seven pairs greatly exceeds the second number, so each of those points falls well below the 45° line. A substantial positive skew (stretched upper tail) is indicated.

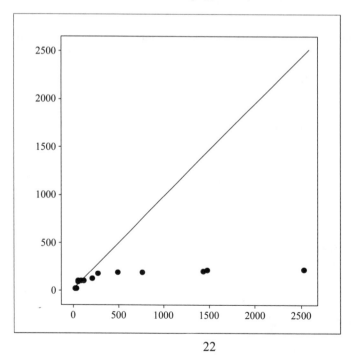

22

CHAPTER 2

Section 2.1

1.

   a. S = {1324, 1342, 1423, 1432, 2314, 2341, 2413, 2431, 3124, 3142, 4123, 4132, 3214, 3241, 4213, 4231}.

   b. Event A contains the outcomes where 1 is first in the list:
A = {1324, 1342, 1423, 1432}.

   c. Event B contains the outcomes where 2 is first or second:
B = {2314, 2341, 2413, 2431, 3214, 3241, 4213, 4231}.

   d. The event $A \cup B$ contains the outcomes in A or B or both:
$A \cup B$ = {1324, 1342, 1423, 1432, 2314, 2341, 2413, 2431, 3214, 3241, 4213, 4231}.
$A \cap B = \varnothing$, since 1 and 2 can't both get into the championship game.
$A' = S - A$ = {2314, 2341, 2413, 2431, 3124, 3142, 4123, 4132, 3214, 3241, 4213, 4231}.

3.

   a. A = {SSF, SFS, FSS}.

   b. B = {SSS, SSF, SFS, FSS}.

   c. For event C to occur, the system must have component 1 working (S in the first position), then at least one of the other two components must work (at least one S in the second and third positions): C = {SSS, SSF, SFS}.

   d. C' = {SFF, FSS, FSF, FFS, FFF}.
$A \cup C$ = {SSS, SSF, SFS, FSS}.
$A \cap C$ = {SSF, SFS}.
$B \cup C$ = {SSS, SSF, SFS, FSS}. Notice that B contains C, so $B \cup C = B$.
$B \cap C$ = {$SSS\ SSF, SFS$}. Since B contains C, $B \cap C = C$.

23

5.

a. The $3^3 = 27$ possible outcomes are numbered below for later reference.

Outcome Number	Outcome	Outcome Number	Outcome
1	111	15	223
2	112	16	231
3	113	17	232
4	121	18	233
5	122	19	311
6	123	20	312
7	131	21	313
8	132	22	321
9	133	23	322
10	211	24	323
11	212	25	331
12	213	26	332
13	221	27	333
14	222		

b. Outcome numbers 1, 14, 27 above.

c. Outcome numbers 6, 8, 12, 16, 20, 22 above.

d. Outcome numbers 1, 3, 7, 9, 19, 21, 25, 27 above.

7.

a. S = {*BBBAAAA, BBABAAA, BBAABAA, BBAAABA, BBAAAAB, BABBAAA, BABABAA, BABAABA, BABAAAB, BAABBAA, BAABABA, BAABAAB, BAAABBA, BAAABAB, BAAAABB, ABBBAAA, ABBABAA, ABBAABA, ABBAAAB, ABABBAA, ABABABA, ABABAAB, ABAABBA, ABAABAB, ABAAABB, AABBBAA, AABBABA, AABBAAB, AABABBA, AABABAB, AABAABB, AAABBBA, AAABBAB, AAABABB, AAAABBB*}.

b. *AAAABBB, AAABABB, AAABBAB, AABAABB, AABABAB.*

9.

a. In the diagram on the left, the shaded area is $(A \cup B)'$. On the right, the shaded area is A', the striped area is B', and the intersection $A' \cap B'$ occurs where there is both shading <u>and</u> stripes. These two diagrams display the same area.

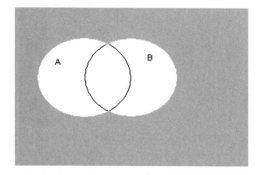   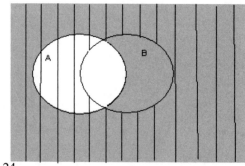

b. In the diagram below, the shaded area represents $(A \cap B)'$. Using the right-hand diagram from (a), the union of A' and B' is represented by the areas that have either shading <u>or</u> stripes (or both). Both of the diagrams display the same area.

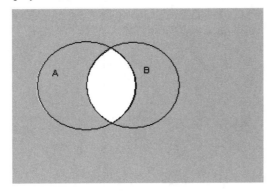

Section 2.2

11.

    a. .07.

    b. $.15 + .10 + .05 = .30$.

    c. Let A = the selected individual owns shares in a stock fund. Then $P(A) = .18 + .25 = .43$. The desired probability, that a selected customer does <u>not</u> shares in a stock fund, equals $P(A') = 1 - P(A) = 1 - .43 = .57$. This could also be calculated by adding the probabilities for all the funds that are not stocks.

13.

    a. $A_1 \cup A_2$ = "awarded either #1 or #2 (or both)": from the addition rule,
$$P(A_1 \cup A_2) = P(A_1) + P(A_2) - P(A_1 \cap A_2) = .22 + .25 - .11 = .36.$$

    b. $A_1' \cap A_2'$ = "awarded neither #1 or #2": using the hint and part (a),
$$P(A_1' \cap A_2') = P((A_1 \cup A_2)') = 1 - P(A_1 \cup A_2) = 1 - .36 = .64.$$

    c. $A_1 \cup A_2 \cup A_3$ = "awarded at least one of these three projects": using the addition rule for 3 events,
$$P(A_1 \cup A_2 \cup A_3) = P(A_1) + P(A_2) + P(A_3) - P(A_1 \cap A_2) - P(A_1 \cap A_3) - P(A_2 \cap A_3) + P(A_1 \cap A_2 \cap A_3) =$$
$$.22 + .25 + .28 - .11 - .05 - .07 + .01 = .53.$$

    d. $A_1' \cap A_2' \cap A_3'$ = "awarded none of the three projects":
$$P(A_1' \cap A_2' \cap A_3') = 1 - P(\text{awarded at least one}) = 1 - .53 = .47.$$

e. $A_1' \cap A_2' \cap A_3$ = "awarded #3 but neither #1 nor #2": from a Venn diagram,

$P(A_1' \cap A_2' \cap A_3) = P(A_3) - P(A_1 \cap A_3) - P(A_2 \cap A_3) + P(A_1 \cap A_2 \cap A_3) =$

$.28 - .05 - .07 + .01 = .17$. The last term addresses the "double counting" of the two subtractions.

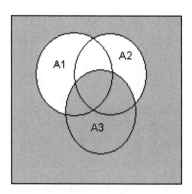

f. $(A_1' \cap A_2') \cup A_3$ = "awarded neither of #1 and #2, or awarded #3": from a Venn diagram,

$P((A_1' \cap A_2') \cup A_3) = P(\text{none awarded}) + P(A_3) = .47$ (from **d**) $+ .28 = 75$.

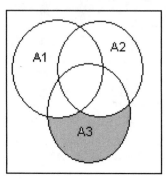

Alternatively, answers to **a-f** can be obtained from probabilities on the accompanying Venn diagram:

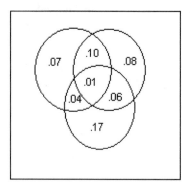

15.

   a. Let E be the event that at most one purchases an electric dryer. Then E' is the event that at least two purchase electric dryers, and $P(E') = 1 - P(E) = 1 - .428 = .572$.

   b. Let A be the event that all five purchase gas, and let B be the event that all five purchase electric. All other possible outcomes are those in which at least one of each type of clothes dryer is purchased. Thus, the desired probability is $1 - [P(A) - P(B)] =$
$1 - [.116 + .005] = .879$.

17.

   a. The probabilities do not add to 1 because there are other software packages besides SPSS and SAS for which requests could be made.

   b. $P(A') = 1 - P(A) = 1 - .30 = .70$.

   c. Since A and B are mutually exclusive events, $P(A \cup B) = P(A) + P(B) = .30 + .50 = .80$.

   d. By deMorgan's law, $P(A' \cap B') = P((A \cup B)') = 1 - P(A \cup B) = 1 - .80 = .20$.
In this example, deMorgan's law says the event "neither A nor B" is the complement of the event "either A or B." (That's true regardless of whether they're mutually exclusive.)

19. Let A be that the selected joint was found defective by inspector A, so $P(A) = \frac{724}{10,000}$. Let B be analogous for inspector B, so $P(B) = \frac{751}{10,000}$. The event "at least one of the inspectors judged a joint to be defective is $A \cup B$, so $P(A \cup B) = \frac{1159}{10,000}$.

   a. By deMorgan's law, $P(\text{neither } A \text{ nor } B) = P(A' \cap B') = 1 - P(A \cup B) = 1 - \frac{1159}{10,000} = \frac{8841}{10,000} = .8841$.

   b. The desired event is $B \cap A'$. From a Venn diagram, we see that $P(B \cap A') = P(B) - P(A \cap B)$. From the addition rule, $P(A \cup B) = P(A) + P(B) - P(A \cap B)$ gives $P(A \cap B) = .0724 + .0751 - .1159 = .0316$. Finally, $P(B \cap A') = P(B) - P(A \cap B) = .0751 - .0316 = .0435$.

21. In what follows, the first letter refers to the auto deductible and the second letter refers to the homeowner's deductible.

   a. $P(MH) = .10$.

   b. $P(\text{low auto deductible}) = P(\{LN, LL, LM, LH\}) = .04 + .06 + .05 + .03 = .18$. Following a similar pattern, $P(\text{low homeowner's deductible}) = .06 + .10 + .03 = .19$.

   c. $P(\text{same deductible for both}) = P(\{LL, MM, HH\}) = .06 + .20 + .15 = .41$.

   d. $P(\text{deductibles are different}) = 1 - P(\text{same deductible for both}) = 1 - .41 = .59$.

   e. $P(\text{at least one low deductible}) = P(\{LN, LL, LM, LH, ML, HL\}) = .04 + .06 + .05 + .03 + .10 + .03 = .31$.

   f. $P(\text{neither deductible is low}) = 1 - P(\text{at least one low deductible}) = 1 - .31 = .69$.

23. Assume that the computers are numbered 1-6 as described and that computers 1 and 2 are the two laptops. There are 15 possible outcomes: (1,2) (1,3) (1,4) (1,5) (1,6) (2,3) (2,4) (2,5) (2,6) (3,4) (3,5) (3,6) (4,5) (4,6) and (5,6).

a. $P(\text{both are laptops}) = P(\{(1,2)\}) = \frac{1}{15} = .067$.

b. $P(\text{both are desktops}) = P(\{(3,4)\,(3,5)\,(3,6)\,(4,5)\,(4,6)\,(5,6)\}) = \frac{6}{15} = .40$.

c. $P(\text{at least one desktop}) = 1 - P(\text{no desktops}) = 1 - P(\text{both are laptops}) = 1 - .067 = .933$.

d. $P(\text{at least one of each type}) = 1 - P(\text{both are the same}) = 1 - [P(\text{both are laptops}) + P(\text{both are desktops})] = 1 - [.067 + .40] = .533$.

25. By rearranging the addition rule, $P(A \cap B) = P(A) + P(B) - P(A \cup B) = .40 + .55 - .63 = .32$. By the same method, $P(A \cap C) = .40 + .70 - .77 = .33$ and $P(B \cap C) = .55 + .70 - .80 = .45$. Finally, rearranging the addition rule for 3 events gives
$P(A \cap B \cap C) = P(A \cup B \cup C) - P(A) - P(B) - P(C) + P(A \cap B) + P(A \cap C) + P(B \cap C) = .85 - .40 - .55 - .70 + .32 + .33 + .45 = .30$.
These probabilities are reflected in the Venn diagram below.

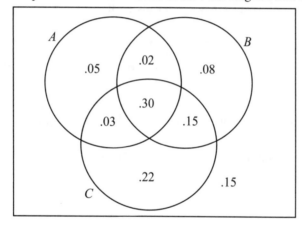

a. $P(A \cup B \cup C) = .85$, as given.

b. $P(\text{none selected}) = 1 - P(\text{at least one selected}) = 1 - P(A \cup B \cup C) = 1 - .85 = .15$.

c. From the Venn diagram, $P(\text{only automatic transmission selected}) = .22$.

d. From the Venn diagram, $P(\text{exactly one of the three}) = .05 + .08 + .22 = .35$.

27. There are 10 equally likely outcomes: {A, B} {A, Co} {A, Cr} {A,F} {B, Co} {B, Cr} {B, F} {Co, Cr} {Co, F} and {Cr, F}.

a. $P(\{A, B\}) = \frac{1}{10} = .1$.

b. $P(\text{at least one C}) = P(\{A, Co\} \text{ or } \{A, Cr\} \text{ or } \{B, Co\} \text{ or } \{B, Cr\} \text{ or } \{Co, Cr\} \text{ or } \{Co, F\} \text{ or } \{Cr, F\}) = \frac{7}{10} = .7$.

c. Replacing each person with his/her years of experience, $P(\text{at least 15 years}) = P(\{3, 14\} \text{ or } \{6, 10\} \text{ or } \{6, 14\} \text{ or } \{7, 10\} \text{ or } \{7, 14\} \text{ or } \{10, 14\}) = \frac{6}{10} = .6$.

Chapter 2: Probability

Section 2.3

29.

a. There are 26 letters, so allowing repeats there are $(26)(26) = (26)^2 = 676$ possible 2-letter domain names. Add in the 10 digits, and there are 36 characters available, so allowing repeats there are $(36)(36) = (36)^2 = 1296$ possible 2-character domain names.

b. By the same logic as part **a**, the answers are $(26)^3 = 17,576$ and $(36)^3 = 46,656$.

c. Continuing, $(26)^4 = 456,976$; $(36)^4 = 1,679,616$.

d. P(4-character sequence is already owned) = 1 – P(4-character sequence still available) = 1 – $97,786/(36)^4 = .942$.

31.

a. Use the Fundamental Counting Principle: $(9)(27) = 243$.

b. By the same reasoning, there are $(9)(27)(15) = 3645$ such sequences, so such a policy could be carried out for 3645 successive nights, or approximately 10 years, without repeating exactly the same program.

33.

a. Since there are 15 players and 9 positions, and order matters in a line-up (catcher, pitcher, shortstop, etc. are different positions), the number of possibilities is $P_{9,15} = (15)(14)...(7)$ or $15!/(15–9)! = 1,816,214,440$.

b. For each of the starting line-ups in part (a), there are 9! possible batting orders. So, multiply the answer from (a) by 9! to get $(1,816,214,440)(362,880) = 659,067,881,472,000$.

c. Order still matters: There are $P_{3,5} = 60$ ways to choose three left-handers for the outfield and $P_{6,10} = 151,200$ ways to choose six right-handers for the other positions. The total number of possibilities is $= (60)(151,200) = 9,072,000$.

35.

a. Since there are 20 day-shift workers, the number of such samples is $\binom{20}{6} = 38,760$. With 45 workers total, there are $\binom{45}{6}$ total possible samples. So, the probability of randomly selecting all day-shift workers is $\dfrac{\binom{20}{6}}{\binom{45}{6}} = \dfrac{38,760}{8,145,060} = .0048$.

b. Following the analogy from **a**, P(all from the same shift) = P(all from day shift) + P(all from swing shift) + P(all from graveyard shift) $= \dfrac{\binom{20}{6}}{\binom{45}{6}} + \dfrac{\binom{15}{6}}{\binom{45}{6}} + \dfrac{\binom{10}{6}}{\binom{45}{6}} = .0048 + .0006 + .0000 = .0054$.

29

© 2012 Cengage Learning. All Rights Reserved. May not be scanned, copied or duplicated, or posted to a publicly accessible website, in whole or in part.

c. $P(\text{at least two shifts represented}) = 1 - P(\text{all from same shift}) = 1 - .0054 = .9946.$

d. There are several ways to approach this question. For example, let A_1 = "day shift is unrepresented," A_2 = "swing shift is unrepresented," and A_3 = "graveyard shift is unrepresented." Then we want $P(A_1 \cup A_2 \cup A_3)$.

$$P(A_1) = P(\text{day shift unrepresented}) = P(\text{all from swing/graveyard}) = \frac{\binom{25}{6}}{\binom{45}{6}},$$

since there are $15 + 10 = 25$ total employees in the swing and graveyard shifts. Similarly,

$$P(A_2) = \frac{\binom{30}{6}}{\binom{45}{6}} \text{ and } P(A_3) = \frac{\binom{35}{6}}{\binom{45}{6}}. \text{ Next, } P(A_1 \cap A_2) = P(\text{all from graveyard}) = \frac{\binom{10}{6}}{\binom{45}{6}}.$$

Similarly, $P(A_1 \cap A_3) = \dfrac{\binom{15}{6}}{\binom{45}{6}}$ and $P(A_2 \cap A_3) = \dfrac{\binom{20}{6}}{\binom{45}{6}}$. Finally, $P(A_1 \cap A_2 \cap A_3) = 0$, since at least one

shift must be represented. Now, apply the addition rule for 3 events:

$$P(A_1 \cup A_2 \cup A_3) = \frac{\binom{25}{6}}{\binom{45}{6}} + \frac{\binom{30}{6}}{\binom{45}{6}} + \frac{\binom{35}{6}}{\binom{45}{6}} - \frac{\binom{10}{6}}{\binom{45}{6}} - \frac{\binom{15}{6}}{\binom{45}{6}} - \frac{\binom{20}{6}}{\binom{45}{6}} + 0 = .2885.$$

37.

a. By the Fundamental Counting Principle, with $n_1 = 3$, $n_2 = 4$, and $n_3 = 5$, there are $(3)(4)(5) = 60$ runs.

b. With $n_1 = 1$ (just one temperature), $n_2 = 2$, and $n_3 = 5$, there are $(1)(2)(5) = 10$ such runs.

c. For each of the 5 specific catalysts, there are $(3)(4) = 12$ pairings of temperature and pressure. Imagine we separate the 60 possible runs into those 5 sets of 12. The number of ways to select exactly one run from each of these 5 sets of 12 is $\binom{12}{1}^5 = 12^5$.

Since there are $\binom{60}{5}$ ways to select the 5 runs overall, the desired probability is $\dfrac{\binom{12}{1}^5}{\binom{60}{5}} = \dfrac{12^5}{\binom{60}{5}} = .0456.$

39.

a. We want to choose all of the 5 cordless, and 5 of the 10 others, to be among the first 10 serviced, so the desired probability is $\dfrac{\binom{5}{5}\binom{10}{5}}{\binom{15}{10}} = \dfrac{252}{3003} = .0839$.

b. Isolating one group, say the cordless phones, we want the other two groups (cellular and corded) represented in the last 5 serviced. The number of ways to choose all 5 cordless phones and 5 of the other phones in the first 10 selections is $\binom{5}{5}\binom{10}{5} = \binom{10}{5}$. However, we don't want <u>two</u> types to be eliminated in the first 10 selections, so we must subtract out the ways that either (all cordless and all cellular) or (all cordless and all corded) are selected among the first 10, which is $\binom{5}{5}\binom{5}{5} + \binom{5}{5}\binom{5}{5} = 2$.

So, the number of ways to have only cellular and corded phones represented in the last five selections is $\binom{10}{5} - 2$. We have three types of phones, so the total number of ways to have exactly two types left over is $3 \cdot \left[\binom{10}{5} - 2\right]$, and the probability is $\dfrac{3 \cdot \left[\binom{10}{5} - 2\right]}{\binom{15}{5}} = \dfrac{3(250)}{3003} = .2498$.

c. We want to choose 2 of the 5 cordless, 2 of the 5 cellular, and 2 of the corded phones:
$\dfrac{\binom{5}{2}\binom{5}{2}\binom{5}{2}}{\binom{15}{6}} = \dfrac{1000}{5005} = .1998$.

41.

a. $(10)(10)(10)(10) = 10^4 = 10,000$. These are the strings 0000 through 9999.

b. Count the number of prohibited sequences. There are (i) 10 with all digits identical (0000, 1111, …, 9999); (ii) 14 with sequential digits (0123, 1234, 2345, 3456, 4567, 5678, 6789, and 7890, plus these same seven descending); (iii) 100 beginning with 19 (1900 through 1999). That's a total of $10 + 14 + 100 = 124$ impermissible sequences, so there are a total of $10,000 - 124 = 9876$ permissible sequences. The chance of randomly selecting one is just $\dfrac{9876}{10,000} = .9876$.

c. All PINs of the form 8xx1 are legitimate, so there are $(10)(10) = 100$ such PINs. With someone randomly selecting 3 such PINs, the chance of guessing the correct sequence is $3/100 = .03$.

d. Of all the PINs of the form 1xx1, eleven is prohibited: 1111, and the ten of the form 19x1. That leaves 89 possibilities, so the chances of correctly guessing the PIN in 3 tries is $3/89 = .0337$.

43.   There are $\binom{52}{5} = 2{,}598{,}960$ five-card hands. The number of 10-high straights is $(4)(4)(4)(4)(4) = 4^5 = 1024$

(any of four 6s, any of four 7s, etc.). So, $P(10 \text{ high straight}) = \dfrac{1024}{2{,}598{,}960} = .000394$. Next, there ten "types

of straight: A2345, 23456, ..., 910JQK, 10JQKA. So, $P(\text{straight}) = 10 \times \dfrac{1024}{2{,}598{,}960} = .00394$. Finally, there

are only 40 straight flushes: each of the ten sequences above in each of the 4 suits makes $(10)(4) = 40$. So,

$P(\text{straight flush}) = \dfrac{40}{2{,}598{,}960} = .00001539$.

Section 2.4

45.

   a. $P(A) = .106 + .141 + .200 = .447$, $P(C) = .215 + .200 + .065 + .020 = .500$, and $P(A \cap C) = .200$.

   b. $P(A/C) = \dfrac{P(A \cap C)}{P(C)} = \dfrac{.200}{.500} = .400$. If we know that the individual came from ethnic group 3, the

   probability that he has Type A blood is .40. $P(C/A) = \dfrac{P(A \cap C)}{P(A)} = \dfrac{.200}{.447} = .447$. If a person has Type A

   blood, the probability that he is from ethnic group 3 is .447.

   c. Define D = "ethnic group 1 selected." We are asked for $P(D/B')$. From the table, $P(D \cap B') = .082 + .106 + .004 = .192$ and $P(B') = 1 - P(B) = 1 - [.008 + .018 + .065] = .909$. So, the desired probability is

   $P(D/B') = \dfrac{P(D \cap B')}{P(B')} = \dfrac{.192}{.909} = .211$.

47.

   a. $P(B/A) = \dfrac{P(A \cap B)}{P(A)} = \dfrac{.25}{.50} = .50$.

   b. $P(B'|A) = \dfrac{P(A \cap B')}{P(A)} = \dfrac{.25}{.50} = .50$.

   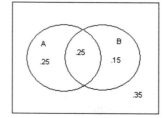

   c. $P(A/B) = \dfrac{P(A \cap B)}{P(B)} = \dfrac{.25}{.40} = .6125$.

   d. $P(A'|B) = \dfrac{P(A' \cap B)}{P(B)} = \dfrac{.15}{.40} = .3875$.

   e. $P(A|A \cup B) = \dfrac{P(A \cap (A \cup B))}{P(A \cup B)} = \dfrac{P(A)}{P(A \cup B)} = \dfrac{.50}{.65} = .7692$.

   It should be clear from the Venn diagram that $A \cap (A \cup B) = A$.

49.

    a. $P(\text{small cup}) = .14 + .20 = .34$. $P(\text{decaf}) = .20 + .10 + .10 = .40$.

    b. $P(\text{decaf} \mid \text{small}) = \dfrac{P(\text{small} \cap \text{decaf})}{P(\text{small})} = \dfrac{.20}{.34} = .588$. 58.8% of all people who purchase a small cup of coffee choose decaf.

    c. $P(\text{small} \mid \text{decaf}) = \dfrac{P(\text{small} \cap \text{decaf})}{P(\text{decaf})} = \dfrac{.20}{.40} = .50$. 50% of all people who purchase decaf coffee choose the small size.

51.

    a. If a red ball is drawn from the first box, the composition of the second box becomes eight red and three green. Use the multiplication rule:

$$P(\text{R from } 1^{\text{st}} \cap \text{R from } 2^{\text{nd}}) = P(\text{R from } 1^{\text{st}}) \times P(\text{R from } 2^{\text{nd}} \mid \text{R from } 1^{\text{st}}) = \frac{6}{10} \times \frac{8}{11} = .436.$$

    b. $P(\text{same numbers as originally}) = P(\text{both selected balls are the same color}) = P(\text{both R}) + P(\text{both G}) =$
$$\frac{6}{10} \times \frac{8}{11} + \frac{4}{10} \times \frac{4}{11} = .581.$$

53. $P(B/A) = \dfrac{P(A \cap B)}{P(A)} = \dfrac{P(B)}{P(A)} = \dfrac{.05}{.60} = .0833$ (since B is contained in A, $A \cap B = B$).

55. Let $A = \{\text{carries Lyme disease}\}$ and $B = \{\text{carries HGE}\}$. We are told $P(A) = .16$, $P(B) = .10$, and $P(A \cap B \mid A \cup B) = .10$. From this last statement and the fact that $A \cap B$ is contained in $A \cup B$,

$.10 = \dfrac{P(A \cap B)}{P(A \cup B)} \Rightarrow P(A \cap B) = .10 P(A \cup B) = .10[P(A) + P(B) - P(A \cap B)] = .10[.10 + .16 - P(A \cap B)] \Rightarrow$

$1.1 P(A \cap B) = .026 \Rightarrow P(A \cap B) = .02364$.

Finally, the desired probability is $P(A \mid B) = \dfrac{P(A \cap B)}{P(B)} = \dfrac{.02364}{.10} = .2364$.

57. $P(B \mid A) > P(B)$ iff $P(B \mid A) + P(B' \mid A) > P(B) + P(B' \mid A)$ iff $1 > P(B) + P(B' \mid A)$ by Exercise 56 (with the letters switched). This holds iff $1 - P(B) > P(B' \mid A)$ iff $P(B') > P(B' \mid A)$, QED.

59. The required probabilities appear in the tree diagram below.

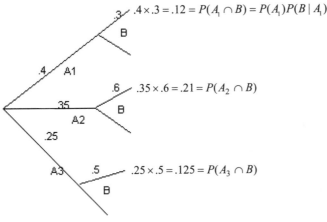

a. $P(A_2 \cap B) = .21$.

b. By the law of total probability, $P(B) = P(A_1 \cap B) + P(A_2 \cap B) + P(A_3 \cap B) = .455$.

c. Using Bayes' theorem, $P(A_1 \mid B) = \dfrac{P(A_1 \cap B)}{P(B)} = \dfrac{.12}{.455} = .264$; $P(A_2 \mid B) = \dfrac{.21}{.455} = .462$; $P(A_3 \mid B) = 1 -$ $.264 - .462 = .274$. Notice the three probabilities sum to 1.

61. The initial ("prior") probabilities of 0, 1, 2 defectives in the batch are .5, .3, .2. Now, let's determine the probabilities of 0, 1, 2 defectives in the sample based on these three cases.
   • If there are 0 defectives in the batch, clearly there are 0 defectives in the sample.
   $P(0 \text{ def in sample} \mid 0 \text{ def in batch}) = 1$.
   • If there is 1 defective in the batch, the chance it's discovered in a sample of 2 equals $2/10 = .2$, and the probability it isn't discovered is $8/10 = .8$.
   $P(0 \text{ def in sample} \mid 1 \text{ def in batch}) = .8$, $P(1 \text{ def in sample} \mid 1 \text{ def in batch}) = .2$.
   • If there are 2 defectives in the batch, the chance both are discovered in a sample of 2 equals $\dfrac{2}{10} \times \dfrac{1}{9} = .022$; the chance neither is discovered equals $\dfrac{8}{10} \times \dfrac{7}{9} = .622$; and the chance exactly 1 is discovered equals $1 - (.022 + .622) = .356$.
   $P(0 \text{ def in sample} \mid 2 \text{ def in batch}) = .622$, $P(1 \text{ def in sample} \mid 2 \text{ def in batch}) = .356$,
   $P(2 \text{ def in sample} \mid 2 \text{ def in batch}) = .022$.

These calculations are summarized in the tree diagram below. Probabilities at the endpoints are intersectional probabilities, e.g. $P(2 \text{ def in batch} \cap 2 \text{ def in sample}) = (.2)(.022) = .0044$.

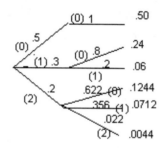

a. Using the tree diagram and Bayes' rule,
$$P(0 \text{ def in batch} \mid 0 \text{ def in sample}) = \frac{.5}{.5 + .24 + .1244} = .578$$
$$P(1 \text{ def in batch} \mid 0 \text{ def in sample}) = \frac{.24}{.5 + .24 + .1244} = .278$$
$$P(2 \text{ def in batch} \mid 0 \text{ def in sample}) = \frac{.1244}{.5 + .24 + .1244} = .144$$

b. $P(0 \text{ def in batch} \mid 1 \text{ def in sample}) = 0$
$$P(1 \text{ def in batch} \mid 1 \text{ def in sample}) = \frac{.06}{.06 + .0712} = .457$$
$$P(2 \text{ def in batch} \mid 1 \text{ def in sample}) = \frac{.0712}{.06 + .0712} = .543$$

63.

  a.

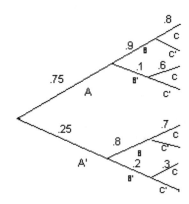

  b. From the top path of the tree diagram, $P(A \cap B \cap C) = (.75)(.9)(.8) = .54$.

  c. Event $B \cap C$ occurs twice on the diagram: $P(B \cap C) = P(A \cap B \cap C) + P(A' \cap B \cap C) = .54 + (.25)(.8)(.7) = .68$.

  d. $P(C) = P(A \cap B \cap C) + P(A' \cap B \cap C) + P(A \cap B' \cap C) + P(A' \cap B' \cap C) = .54 + .045 + .14 + .015 = .74$.

  e. Rewrite the conditional probability first: $P(A \mid B \cap C) = \dfrac{P(A \cap B \cap C)}{P(B \cap C)} = \dfrac{.54}{.68} = .7941$.

65. A tree diagram can help. We know that $P(\text{day}) = .2$, $P(\text{1-night}) = .5$, $P(\text{2-night}) = .3$; also, $P(\text{purchase} \mid \text{day}) = .1$, $P(\text{purchase} \mid \text{1-night}) = .3$, and $P(\text{purchase} \mid \text{2-night}) = .2$.

Apply Bayes' rule: e.g., $P(\text{day} \mid \text{purchase}) = \dfrac{P(\text{day} \cap \text{purchase})}{P(\text{purchase})} = \dfrac{(.2)(.1)}{(.2)(.1) + (.5)(.3) + (.3)(.2)} = \dfrac{.02}{.23} = .087$.

Similarly, $P(\text{1-night} \mid \text{purchase}) = \dfrac{(.5)(.3)}{.23} = .652$ and $P(\text{2-night} \mid \text{purchase}) = .261$.

67. Let T denote the event that a randomly selected person is, in fact, a terrorist. Apply Bayes' theorem, using $P(T) = 1{,}000/300{,}000{,}000 = .0000033$:

$$P(T \mid +) = \frac{P(T)P(+ \mid T)}{P(T)P(+ \mid T) + P(T')P(+ \mid T')} = \frac{(.0000033)(.99)}{(.0000033)(.99) + (1 - .0000033)(1 - .999)} = .003289. \text{ That is to}$$

say, roughly 0.3% of all people "flagged" as terrorists would be actual terrorists in this scenario.

69. The tree diagram below summarizes the information in the exercise (plus the previous information in Exercise 59). Probabilities for the branches corresponding to paying with credit are indicated at the far right. ("extra" = "plus")

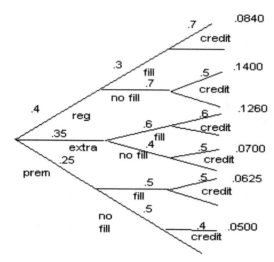

a. $P(\text{plus} \cap \text{fill} \cap \text{credit}) = (.35)(.6)(.6) = .1260$.

b. $P(\text{premium} \cap \text{no fill} \cap \text{credit}) = (.25)(.5)(.4) = .05$.

c. From the tree diagram, $P(\text{premium} \cap \text{credit}) = .0625 + .0500 = .1125$.

d. From the tree diagram, $P(\text{fill} \cap \text{credit}) = .0840 + .1260 + .0625 = .2725$.

e. $P(\text{credit}) = .0840 + .1400 + .1260 + .0700 + .0625 + .0500 = .5325$.

f. $P(\text{premium} \mid \text{credit}) = \dfrac{P(\text{premium} \cap \text{credit})}{P(\text{credit})} = \dfrac{.1125}{.5325} = .2113$.

Section 2.5

71.

a. Since the events are independent, then A' and B' are independent, too. (See the paragraph below Equation 2.7.) Thus, $P(B'|A') = P(B') = 1 - .7 = .3$.

b. Using the addition rule, $P(A \cup B) = P(A) + P(B) - P(A \cap B) = .4 + .7 - (.4)(.7) = .82$. Since A and B are independent, we are permitted to write $P(A \cap B) = P(A)P(B) = (.4)(.7)$.

c. $P(AB' \mid A \cup B) = \dfrac{P(AB' \cap (A \cup B))}{P(A \cup B)} = \dfrac{P(AB')}{P(A \cup B)} = \dfrac{P(A)P(B')}{P(A \cup B)} = \dfrac{(.4)(1-.7)}{.82} = \dfrac{.12}{.82} = .146$.

73. From a Venn diagram, $P(B) = P(A' \cap B) + P(A \cap B) = P(B) \Rightarrow P(A' \cap B) = P(B) - P(A \cap B)$. If A and B are independent, then $P(A' \cap B) = P(B) - P(A)P(B) = [1 - P(A)]P(B) = P(A')P(B)$. Thus, A' and B are independent.

Alternatively, $P(A' \mid B) = \dfrac{P(A' \cap B)}{P(B)} = \dfrac{P(B) - P(A \cap B)}{P(B)} = \dfrac{P(B) - P(A)P(B)}{P(B)} = 1 - P(A) = P(A')$.

75. Let event E be the event that an error was signaled incorrectly.

We want P(at least one signaled incorrectly) $= P(E_1 \cup \ldots \cup E_{10})$. To use independence, we need intersections, so apply deMorgan's law: $= P(E_1 \cup \ldots \cup E_{10}) = 1 - P(E_1' \cap \cdots \cap E_{10}')$. $P(E') = 1 - .05 = .95$, so for 10 independent points, $P(E_1' \cap \cdots \cap E_{10}') = (.95)\ldots(.95) = (.95)^{10}$. Finally, $P(E_1 \cup E_2 \cup \ldots \cup E_{10}) = 1 - (.95)^{10} = .401$. Similarly, for 25 points, the desired probability is $1 - (P(E'))^{25} = 1 - (.95)^{25} = .723$.

77. Let p denote the probability that a rivet is defective.

a. $.20 = P$(seam needs reworking) $= 1 - P$(seam doesn't need reworking) $=$
$1 - P$(no rivets are defective) $= 1 - P(1^{st}$ isn't def $\cap \ldots \cap 25^{th}$ isn't def$) =$
$1 - (1 - p)\ldots(1 - p) = 1 - (1 - p)^{25}$.
Solve for p: $(1 - p)^{25} = .80 \Rightarrow 1 - p = (.80)^{1/25} \Rightarrow p = 1 - .99111 = .00889$.

b. The desired condition is $.10 = 1 - (1 - p)^{25}$. Again, solve for p: $(1 - p)^{25} = .90 \Rightarrow$
$p = 1 - (.90)^{1/25} = 1 - .99579 = .00421$.

79. Let $A_1 =$ older pump fails, $A_2 =$ newer pump fails, and $x = P(A_1 \cap A_2)$. The goal is to find x. From the Venn diagram below, $P(A_1) = .10 + x$ and $P(A_2) = .05 + x$. Independence implies that $x = P(A_1 \cap A_2) = P(A_1)P(A_2)$ $= (.10 + x)(.05 + x)$. The resulting quadratic equation, $x^2 - .85x + .005 = 0$, has roots $x = .0059$ and $x = .8441$. The latter is impossible, since the probabilities in the Venn diagram would then exceed 1. Therefore, $x = .0059$.

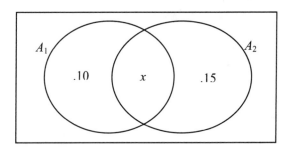

81. Using the hints, let $P(A_i) = p$, and $x = p^2$. Following the solution provided in the example, P(system lifetime exceeds t_0) $= p^2 + p^2 - p^4 = 2p^2 - p^4 = 2x - x^2$. Now, set this equal to .99: $2x - x^2 = .99 \Rightarrow x^2 - 2x + .99 = 0 \Rightarrow x = 0.9$ or $1.1 \Rightarrow p = 1.049$ or $.9487$. Since the value we want is a probability and cannot exceed 1, the correct answer is $p = .9487$.

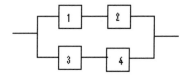

83. We'll need to know P(both detect the defect) = $1 - P$(at least one doesn't) = $1 - .2 = .8$.

a. $P(1^{st}$ detects $\cap$ 2nd doesn't) = $P(1^{st}$ detects) – $P(1^{st}$ does $\cap$ 2nd does) = $.9 - .8 = .1$.
Similarly, $P(1^{st}$ doesn't $\cap$ 2nd does) = $.1$, so P(exactly one does) = $.1 + .1 = .2$.

b. P(neither detects a defect) = $1 - [P$(both do) + P(exactly 1 does)] = $1 - [.8+.2] = 0$. That is, under this model there is a 0% probability neither inspector detects a defect. As a result, P(all 3 escape) = $(0)(0)(0) = 0$.

85.

a. Let D_1 = detection on 1st fixation, D_2 = detection on 2nd fixation.
P(detection in at most 2 fixations) = $P(D_1) + P(D_1' \cap D_2)$; since the fixations are independent,
$P(D_1) + P(D_1' \cap D_2) = P(D_1) + P(D_1')\,P(D_2) = p + (1-p)p = p(2-p)$.

b. Define $D_1, D_2, \ldots, D_n$ as in **a**. Then P(at most n fixations) =
$P(D_1) + P(D_1' \cap D_2) + P(D_1' \cap D_2' \cap D_3) + \ldots + P(D_1' \cap D_2' \cap \cdots \cap D_{n-1}' \cap D_n) =$
$p + (1-p)p + (1-p)^2 p + \ldots + (1-p)^{n-1}p = p[1 + (1-p) + (1-p)^2 + \ldots + (1-p)^{n-1}] =$
$p \cdot \dfrac{1-(1-p)^n}{1-(1-p)} = 1-(1-p)^n$.

Alternatively, P(at most n fixations) = $1 - P$(at least $n+1$ fixations are required) =
$1 - P$(no detection in 1st n fixations) = $1 - P(D_1' \cap D_2' \cap \cdots \cap D_n') = 1 - (1-p)^n$.

c. P(no detection in 3 fixations) = $(1-p)^3$.

d. P(passes inspection) = $P(\{$not flawed$\} \cup \{$flawed and passes$\})$
$= P$(not flawed) + P(flawed and passes)
$= .9 + P$(flawed) P(passes | flawed) = $.9 + (.1)(1-p)^3$.

e. Borrowing from **d**, P(flawed | passed) = $\dfrac{P(\text{flawed} \cap \text{passed})}{P(\text{passed})} = \dfrac{.1(1-p)^3}{.9+.1(1-p)^3}$. For $p = .5$,

P(flawed | passed) = $\dfrac{.1(1-.5)^3}{.9+.1(1-.5)^3} = .0137$.

87.

a. Use the information provided and the addition rule:
$P(A_1 \cup A_2) = P(A_1) + P(A_2) - P(A_1 \cap A_2) \Rightarrow P(A_1 \cap A_2) = P(A_1) + P(A_2) - P(A_1 \cup A_2) = .55 + .65 - .80$
$= .40$.

b. By definition, $P(A_2 \mid A_3) = \dfrac{P(A_2 \cap A_3)}{P(A_3)} = \dfrac{.40}{.70} = .5714$. If a person likes vehicle #3, there's a 57.14% chance s/he will also like vehicle #2.

c. No. From **b**, $P(A_2 \mid A_3) = .5714 \neq P(A_2) = .65$. Therefore, A_2 and A_3 are not independent. Alternatively, $P(A_2 \cap A_3) = .40 \neq P(A_2)P(A_3) = (.65)(.70) = .455$.

d. The goal is to find $P(A_2 \cup A_3 \mid A_1')$, i.e. $\dfrac{P([A_2 \cup A_3] \cap A_1')}{P(A_1')}$. The denominator is simply $1 - .55 = .45$.

There are several ways to calculate the numerator; the simplest approach using the information provided is to draw a Venn diagram and observe that $P([A_2 \cup A_3] \cap A_1') = P(A_1 \cup A_2 \cup A_3) - P(A_1) = .88 - .55 = .33$. Hence, $P(A_2 \cup A_3 \mid A_1') = \dfrac{.33}{.45} = .7333$.

89. The question asks for $P(\underline{\text{exactly}}$ one tag lost $\mid$ at $\underline{\text{most}}$ one tag lost$) = P((C_1 \cap C_2') \cup (C_1' \cap C_2) \mid (C_1 \cap C_2)')$.
Since the first event is contained in (a subset of) the second event, this equals
$$\frac{P((C_1 \cap C_2') \cup (C_1' \cap C_2))}{P((C_1 \cap C_2)')} = \frac{P(C_1 \cap C_2') + P(C_1' \cap C_2)}{1 - P(C_1 \cap C_2)} = \frac{P(C_1)P(C_2') + P(C_1')P(C_2)}{1 - P(C_1)P(C_2)} \text{ by independence} =$$
$$\frac{\pi(1-\pi) + (1-\pi)\pi}{1 - \pi^2} = \frac{2\pi(1-\pi)}{1-\pi^2} = \frac{2\pi}{1+\pi}.$$

Supplementary Exercises

91.

a. $P(\text{line 1}) = \dfrac{500}{1500} = .333$;

$P(\text{crack}) = \dfrac{.50(500) + .44(400) + .40(600)}{1500} = \dfrac{666}{1500} = .444.$

b. This is one of the percentages provided: $P(\text{blemish} \mid \text{line 1}) = .15$.

c. $P(\text{surface defect}) = \dfrac{.10(500) + .08(400) + .15(600)}{1500} = \dfrac{172}{1500}$;

$P(\text{line 1} \cap \text{surface defect}) = \dfrac{.10(500)}{1500} = \dfrac{50}{1500}$;

so, $P(\text{line 1} \mid \text{surface defect}) = \dfrac{50/1500}{172/1500} = \dfrac{50}{172} = .291.$

93. Apply the addition rule: $P(A \cup B) = P(A) + P(B) - P(A \cap B) \Rightarrow .626 = P(A) + P(B) - .144$. Apply independence: $P(A \cap B) = P(A)P(B) = .144$.
So, $P(A) + P(B) = .770$ and $P(A)P(B) = .144$.
Let $x = P(A)$ and $y = P(B)$. Using the first equation, $y = .77 - x$, and substituting this into the second equation yields $x(.77 - x) = .144$ or $x^2 - .77x + .144 = 0$. Use the quadratic formula to solve:
$$x = \frac{.77 \pm \sqrt{(-.77)^2 - (4)(1)(.144)}}{2(1)} = \frac{.77 \pm .13}{2} = .32 \text{ or } .45. \text{ Since } x = P(A) \text{ is assumed to be the larger}$$
probability, $x = P(A) = .45$ and $y = P(B) = .32$.

95.

a. There are 5! = 120 possible orderings, so $P(\text{BCDEF}) = \frac{1}{120} = .0833$.

b. The number of orderings in which F is third equals $4 \times 3 \times 1 * \times 2 \times 1 = 24$ (*because F must be here), so $P(\text{F is third}) = \frac{24}{120} = .2$. Or more simply, since the five friends are ordered completely at random, there is a ⅕ chance F is specifically in position three.

c. Similarly, $P(\text{F last}) = \dfrac{4 \times 3 \times 2 \times 1 \times 1}{120} = .2$.

d. $P(\text{F hasn't heard after 10 times}) = P(\text{not on \#1} \cap \text{not on \#2} \cap \ldots \cap \text{not on \#10}) = \dfrac{4}{5} \times \cdots \times \dfrac{4}{5} = \left(\dfrac{4}{5}\right)^{10} = .1074$.

97. When three experiments are performed, there are 3 different ways in which detection can occur on exactly 2 of the experiments: (i) #1 and #2 and not #3; (ii) #1 and not #2 and #3; and (iii) not #1 and #2 and #3. If the impurity is present, the probability of exactly 2 detections in three (independent) experiments is $(.8)(.8)(.2) + (.8)(.2)(.8) + (.2)(.8)(.8) = .384$. If the impurity is absent, the analogous probability is $3(.1)(.1)(.9) = .027$. Thus, applying Bayes' theorem, $P(\text{impurity is present} \mid \text{detected in exactly 2 out of 3})$

$= \dfrac{P(\text{detected in exactly 2} \cap \text{present})}{P(\text{detected in exactly 2})} = \dfrac{(.384)(.4)}{(.384)(.4) + (.027)(.6)} = .905$.

99. Refer to the tree diagram below.

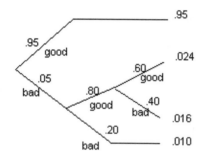

a. $P(\text{pass inspection}) = P(\text{pass initially} \cup \text{passes after recrimping}) =$
$P(\text{pass initially}) + P(\text{fails initially} \cap \text{goes to recrimping} \cap \text{is corrected after recrimping}) =$
$.95 + (.05)(.80)(.60)$ (following path "bad-good-good" on tree diagram) $= .974$.

b. $P(\text{needed no recrimping} \mid \text{passed inspection}) = \dfrac{P(\text{passed initially})}{P(\text{passed inspection})} = \dfrac{.95}{.974} = .9754$.

101. Let $A = 1^{\text{st}}$ functions, $B = 2^{\text{nd}}$ functions, so $P(B) = .9$, $P(A \cup B) = .96$, $P(A \cap B) = .75$. Use the addition rule:
$P(A \cup B) = P(A) + P(B) - P(A \cap B) \Rightarrow .96 = P(A) + .9 - .75 \Rightarrow P(A) = .81$.
Therefore, $P(B \mid A) = \dfrac{P(B \cap A)}{P(A)} = \dfrac{.75}{.81} = .926$.

103. A tree diagram can also help here.

    a. The law of total probability gives $P(L) = \sum P(E_i)P(L \mid E_i) = (.40)(.02) + (.50)(.01) + (.10)(.05) = .018$.

    b. $P(E_1' \mid L') = 1 - P(E_1 \mid L') = 1 - \dfrac{P(E_1 \cap L')}{P(L')} = 1 - \dfrac{P(E_1)P(L' \mid E_1)}{1 - P(L)} = 1 - \dfrac{(.40)(.98)}{1 - .018} = .601$.

105. This is the famous "Birthday Problem" in probability.

    a. There are 365^{10} possible lists of birthdays, e.g. (Dec 10, Sep 27, Apr 1, …). Among those, the number with zero matching birthdays is $P_{10,365}$ (sampling ten birthdays without replacement from 365 days. So,

$$P(\text{all different}) = \frac{P_{10,365}}{365^{10}} = \frac{(365)(364)\cdots(356)}{(365)^{10}} = .883. \; P(\text{at least two the same}) = 1 - .883 = .117.$$

    b. The general formula is $P(\text{at least two the same}) = 1 - \dfrac{P_{k,365}}{365^k}$. By trial and error, this probability equals .476 for $k = 22$ and equals .507 for $k = 23$. Therefore, the smallest k for which k people have at least a 50-50 chance of a birthday match is 23.

    c. There are 1000 possible 3-digit sequences to end a SS number (000 through 999). Using the idea from **a**, $P(\text{at least two have the same SS ending}) = 1 - \dfrac{P_{10,1000}}{1000^{10}} = 1 - .956 = .044$.

    Assuming birthdays and SS endings are independent, $P(\text{at least one "coincidence"}) = P(\text{birthday coincidence} \cup \text{SS coincidence}) = .117 + .044 - (.117)(.044) = .156$.

107. $P(\text{detection by the end of the } n\text{th glimpse}) = 1 - P(\text{not detected in first } n \text{ glimpses}) =$

$$1 - P(G_1' \cap G_2' \cap \cdots \cap G_n') = 1 - P(G_1')P(G_2')\cdots P(G_n') = 1 - (1 - p_1)(1 - p_2)\ldots(1 - p_n) = 1 - \prod_{i=1}^{n}(1 - p_i).$$

109.

    a. $P(\text{all in correct room}) = \dfrac{1}{4!} = \dfrac{1}{24} = .0417$.

    b. The 9 outcomes which yield completely incorrect assignments are: 2143, 2341, 2413, 3142, 3412, 3421, 4123, 4321, and 4312, so $P(\text{all incorrect}) = \dfrac{9}{24} = .375$.

111. Note: $s = 0$ means that the very first candidate interviewed is hired. Each entry below is the candidate hired for the given policy and outcome.

Outcome	$s=0$	$s=1$	$s=2$	$s=3$	Outcome	$s=0$	$s=1$	$s=2$	$s=3$
1234	1	4	4	4	3124	3	1	4	4
1243	1	3	3	3	3142	3	1	4	2
1324	1	4	4	4	3214	3	2	1	4
1342	1	2	2	2	3241	3	2	1	1
1423	1	3	3	3	3412	3	1	1	2
1432	1	2	2	2	3421	3	2	2	1
2134	2	1	4	4	4123	4	1	3	3
2143	2	1	3	3	4132	4	1	2	2
2314	2	1	1	4	4213	4	2	1	3
2341	2	1	1	1	4231	4	2	1	1
2413	2	1	1	3	4312	4	3	1	2
2431	2	1	1	1	4321	4	3	2	1

From the table, we derive the following probability distribution based on s:

s	0	1	2	3
P(hire #1)	$\dfrac{6}{24}$	$\dfrac{11}{24}$	$\dfrac{10}{24}$	$\dfrac{6}{24}$

Therefore $s = 1$ is the best policy.

113. $P(A_1) = P(\text{draw slip 1 or 4}) = \frac{1}{2}$; $P(A_2) = P(\text{draw slip 2 or 4}) = \frac{1}{2}$;

$P(A_3) = P(\text{draw slip 3 or 4}) = \frac{1}{2}$; $P(A_1 \cap A_2) = P(\text{draw slip 4}) = \frac{1}{4}$;

$P(A_2 \cap A_3) = P(\text{draw slip 4}) = \frac{1}{4}$; $P(A_1 \cap A_3) = P(\text{draw slip 4}) = \frac{1}{4}$.

Hence $P(A_1 \cap A_2) = P(A_1)P(A_2) = \frac{1}{4}$; $P(A_2 \cap A_3) = P(A_2)P(A_3) = \frac{1}{4}$; and

$P(A_1 \cap A_3) = P(A_1)P(A_3) = \frac{1}{4}$. Thus, there exists pairwise independence. However,

$P(A_1 \cap A_2 \cap A_3) = P(\text{draw slip 4}) = \frac{1}{4} \neq \frac{1}{8} = P(A_1)P(A_2)P(A_3)$, so the events are not mutually independent.

CHAPTER 3

Section 3.1

1.

S:	FFF	SFF	FSF	FFS	FSS	SFS	SSF	SSS
X:	0	1	1	1	2	2	2	3

3.   Examples include: M = the difference between the large and the smaller outcome with possible values 0, 1, 2, 3, 4, or 5; $T = 1$ if the sum of the two resulting numbers is even and $T = 0$ otherwise, a Bernoulli random variable. See the back of the book for other examples.

5.   No. In the experiment in which a coin is tossed repeatedly until a H results, let $Y = 1$ if the experiment terminates with at most 5 tosses and $Y = 0$ otherwise. The sample space is infinite, yet Y has only two possible values. See the back of the book for another example.

7.

    a.   Possible values of X are 0, 1, 2, …, 12; discrete.

    b.   With n = # on the list, values of Y are 0, 1, 2, … , N; discrete.

    c.   Possible values of U are 1, 2, 3, 4, … ; discrete.

    d.   Possible values of X are $(0, \infty)$ if we assume that a rattlesnake can be arbitrarily short or long; not discrete.

    e.   With c = amount earned in royalties per book sold, possible values of Z are 0, c, $2c$, $3c$, … , 10,000c; discrete.

    f.   Since 0 is the smallest possible pH and 14 is the largest possible pH, possible values of Y are [0, 14]; not discrete.

    g.   With m and M denoting the minimum and maximum possible tension, respectively, possible values of X are [m, M]; not discrete.

    h.   The number of possible tries is 1, 2, 3, …; each try involves 3 coins, so possible values of X are 3, 6, 9, 12, 15, …; discrete.

9.

    a.   Returns to 0 can occur only after an even number of tosses, so possible X values are 2, 4, 6, 8, …. Because the values of X are enumerable, X is discrete.

    b.   Now a return to 0 is possible after any number of tosses greater than 1, so possible values are 2, 3, 4, 5, …. Again, X is discrete.

43

Section 3.2

11.

    a. As displayed in the chart, $p(4) = .45$, $p(6) = .40$, $p(8) = .15$, and $p(x) = 0$ otherwise.

x	4	6	8
$p(x)$	.45	.40	.15

    b.

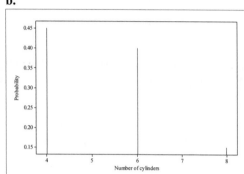 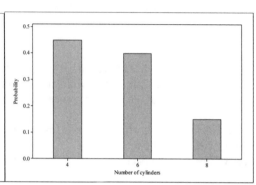

    c. $P(X \geq 6) = .40 + .15 = .55$; $P(X > 6) = P(X = 8) = .15$.

13.

    a. $P(X \leq 3) = p(0) + p(1) + p(2) + p(3) = .10 + .15 + .20 + .25 = .70$.

    b. $P(X < 3) = P(X \leq 2) = p(0) + p(1) + p(2) = .45$.

    c. $P(X \geq 3) = p(3) + p(4) + p(5) + p(6) = .55$.

    d. $P(2 \leq X \leq 5) = p(2) + p(3) + p(4) + p(5) = .71$.

    e. The number of lines <u>not</u> in use is $6 - X$, and $P(2 \leq 6 - X \leq 4) = P(-4 \leq -X \leq -2) =$
    $P(2 \leq X \leq 4) = p(2) + p(3) + p(4) = .65$.

    f. $P(6 - X \geq 4) = P(X \leq 2) = .10 + .15 + .20 = .45$.

15.

    a. (1,2) (1,3) (1,4) (1,5) (2,3) (2,4) (2,5) (3,4) (3,5) (4,5)

    b. X can only take on the values 0, 1, 2. $p(0) = P(X = 0) = P(\{(3,4)\ (3,5)\ (4,5)\}) = 3/10 = .3$;
    $p(2) = P(X = 2) = P(\{(1,2)\}) = 1/10 = .1$; $p(1) = P(X = 1) = 1 - [p(0) + p(2)] = .60$; and otherwise $p(x)$
    $= 0$.

    c. $F(0) = P(X \leq 0) = P(X = 0) = .30$;
    $F(1) = P(X \leq 1) = P(X = 0 \text{ or } 1) = .30 + .60 = .90$;
    $F(2) = P(X \leq 2) = 1$.
    Therefore, the complete cdf of X is

$$F(x) = \begin{cases} 0 & x < 0 \\ .30 & 0 \le x < 1 \\ .90 & 1 \le x < 2 \\ 1 & 2 \le x \end{cases}$$

17.

a. $p(2) = P(Y = 2) = P(\text{first 2 batteries are acceptable}) = P(AA) = (.9)(.9) = .81$.

b. $p(3) = P(Y = 3) = P(UAA \text{ or } AUA) = (.1)(.9)^2 + (.1)(.9)^2 = 2[(.1)(.9)^2] = .162$.

c. The fifth battery must be an A, and exactly one of the first four must also be an A.
Thus, $p(5) = P(AUUUA \text{ or } UAUUA \text{ or } UUAUA \text{ or } UUUAA) = 4[(.1)^3(.9)^2] = .00324$.

d. $p(y) = P(\text{the } y^{th} \text{ is an } A \text{ and so is exactly one of the first } y - 1) = (y - 1)(.1)^{y-2}(.9)^2$, for $y = 2, 3, 4, 5, \ldots$.

19. $p(0) = P(Y = 0) = P(\text{both arrive on Wed}) = (.3)(.3) = .09$;
$p(1) = P(Y = 1) = P((W,Th) \text{ or } (Th,W) \text{ or } (Th,Th)) = (.3)(.4) + (.4)(.3) + (.4)(.4) = .40$;
$p(2) = P(Y = 2) = P((W,F) \text{ or } (Th,F) \text{ or } (F,W) \text{ or } (F,Th) \text{ or } (F,F)) = .32$;
$p(3) = 1 - [.09 + .40 + .32] = .19$.

21.

a. First, $1 + 1/x > 1$ for all $x = 1, \ldots, 9$, so $\log(1 + 1/x) > 0$. Next, check that the probabilities sum to 1:

$$\sum_{x=1}^{9} \log_{10}(1 + 1/x) = \sum_{x=1}^{9} \log_{10}\left(\frac{x+1}{x}\right) = \log_{10}\left(\frac{2}{1}\right) + \log_{10}\left(\frac{3}{2}\right) + \cdots + \log_{10}\left(\frac{10}{9}\right); \text{ using properties of logs,}$$

this equals $\log_{10}\left(\frac{2}{1} \times \frac{3}{2} \times \cdots \times \frac{10}{9}\right) = \log_{10}(10) = 1$.

b. Using the formula $p(x) = \log_{10}(1 + 1/x)$ gives the following values: $p(1) = .301, p(2) = .176, p(3) = .125, p(4) = .097, p(5) = .079, p(6) = .067, p(7) = .058, p(8) = .051, p(9) = .046$. The distribution specified by *Benford's Law* is <u>not</u> uniform on these nine digits; rather, lower digits (such as 1 and 2) are much more likely to be the lead digit of a number than higher digits (such as 8 and 9).

c. The jumps in $F(x)$ occur at $0, \ldots, 8$. We display the cumulative probabilities here: $F(1) = .301, F(2) = .477, F(3) = .602, F(4) = .699, F(5) = .778, F(6) = .845, F(7) = .903, F(8) = .954, F(9) = 1$. So, $F(x) = 0$ for $x < 1$; $F(x) = .301$ for $1 \le x < 2$;
$F(x) = .477$ for $2 \le x < 3$; etc.

d. $P(X \le 3) = F(3) = .602$; $P(X \ge 5) = 1 - P(X < 5) = 1 - P(X \le 4) = 1 - F(4) = 1 - .699 = .301$.

23.

a. $p(2) = P(X = 2) = F(3) - F(2) = .39 - .19 = .20$.

b. $P(X > 3) = 1 - P(X \le 3) = 1 - F(3) = 1 - .67 = .33$.

c. $P(2 \le X \le 5) = F(5) - F(2-1) = F(5) - F(1) = .92 - .19 = .78$.

d. $P(2 < X < 5) = P(2 < X \le 4) = F(4) - F(2) = .92 - .39 = .53$.

25. $p(0) = P(Y = 0) = P(B \text{ first}) = p$;
$p(1) = P(Y = 1) = P(G \text{ first, then } B) = (1 - p)p$;
$p(2) = P(Y = 2) = P(GGB) = (1 - p)^2 p$;
Continuing, $p(y) = P(y \text{ Gs and then a } B) = (1 - p)^y p$ for $y = 0,1,2,3,\ldots$.

27.

   a.   The sample space consists of all possible permutations of the four numbers 1, 2, 3, 4:

outcome	x value	outcome	x value	outcome	x value
1234	4	2314	1	3412	0
1243	2	2341	0	3421	0
1324	2	2413	0	4132	1
1342	1	2431	1	4123	0
1423	1	3124	1	4213	1
1432	2	3142	0	4231	2
2134	2	3214	2	4312	0
2143	0	3241	1	4321	0

   b.   From the table in **a**, $p(0) = P(X = 0) = \frac{9}{24}$, $p(1) = P(X = 1) = \frac{8}{24}$, $p(2) = P(Y = 2) = \frac{6}{24}$, $p(3) = P(X = 3) = 0$, and $p(4) = P(Y = 4) = \frac{1}{24}$.

Section 3.3

29.

   a. $E(X) = \sum_{\text{all } x} x p(x) = 1(.05) + 2(.10) + 4(.35) + 8(.40) + 16(.10) = 6.45$ GB.

   b. $V(X) = \sum_{\text{all } x} (x - \mu)^2 p(x) = (1 - 6.45)^2(.05) + (2 - 6.45)^2(.10) + \ldots + (16 - 6.45)^2(.10) = 15.6475$.

   c. $\sigma = \sqrt{V(X)} = \sqrt{15.6475} = 3.956$ GB.

   d. $E(X^2) = \sum_{\text{all } x} x^2 p(x) = 1^2(.05) + 2^2(.10) + 4^2(.35) + 8^2(.40) + 16^2(.10) = 57.25$. Using the shortcut formula, $V(X) = E(X^2) - \mu^2 = 57.25 - (6.45)^2 = 15.6475$.

31.   From the table in Exercise 12, $E(Y) = 45(.05) + 46(.10) + \ldots + 55(.01) = 48.84$; similarly, $E(Y^2) = 45^2(.05) + 46^2(.10) + \ldots + 55^2(.01) = 2389.84$; thus $V(Y) = E(Y^2) - [E(Y)]^2 = 2389.84 - (48.84)^2 = 4.4944$ and $\sigma_Y = \sqrt{4.4944} = 2.12$.
One standard deviation from the mean value of Y gives $48.84 \pm 2.12 = 46.72$ to 50.96. So, the probability Y is within one standard deviation of its mean value equals $P(46.72 < Y < 50.96) = P(Y = 47, 48, 49, 50) = .12 + .14 + .25 + .17 = .68$.

33.

    a. $E(X^2) = \sum_{x=0}^{1} x^2 \cdot p(x) = 0^2(1-p) + 1^2(p) = p.$

    b. $V(X) = E(X^2) - [E(X)]^2 = p - [p]^2 = p(1-p).$

    c. $E(X^{79}) = 0^{79}(1-p) + 1^{79}(p) = p.$ In fact, $E(X^n) = p$ for any non-negative power n.

35.   Let $h_3(X)$ and $h_4(X)$ equal the net revenue (sales revenue minus order cost) for 3 and 4 copies purchased, respectively. If 3 magazines are ordered ($6 spent), net revenue is $4 – $6 = –$2 if $X = 1$, 2($4) – $6 = $2 if $X = 2$, 3($4) – $6 = $6 if $X = 3$, and also $6 if $X = 4$, 5, or 6 (since that additional demand simply isn't met. The values of $h_4(X)$ can be deduced similarly. Both distributions are summarized below.

x	1	2	3	4	5	6
$h_3(x)$	–2	2	6	6	6	6
$h_4(x)$	–4	0	4	8	8	8
$p(x)$	$\frac{1}{15}$	$\frac{2}{15}$	$\frac{3}{15}$	$\frac{4}{15}$	$\frac{3}{15}$	$\frac{2}{15}$

Using the table, $E[h_3(X)] = \sum_{x=1}^{6} h_3(x) \cdot p(x) = (-2)(\frac{1}{15}) + \ldots + (6)(\frac{2}{15}) = \$4.93.$

Similarly, $E[h_4(X)] = \sum_{x=1}^{6} h_4(x) \cdot p(x) = (-4)(\frac{1}{15}) + \ldots + (8)(\frac{2}{15}) = \$5.33.$

Therefore, ordering 4 copies gives slightly higher revenue, on the average.

37.   Using the hint, $E(X) = \sum_{x=1}^{n} x \cdot \left(\frac{1}{n}\right) = \frac{1}{n}\sum_{x=1}^{n} x = \frac{1}{n}\left[\frac{n(n+1)}{2}\right] = \frac{n+1}{2}.$ Similarly,

$E(X^2) = \sum_{x=1}^{n} x^2 \cdot \left(\frac{1}{n}\right) = \frac{1}{n}\sum_{x=1}^{n} x^2 = \frac{1}{n}\left[\frac{n(n+1)(2n+1)}{6}\right] = \frac{(n+1)(2n+1)}{6}$, so

$V(X) = \frac{(n+1)(2n+1)}{6} - \left(\frac{n+1}{2}\right)^2 = \frac{n^2-1}{12}.$

39.   From the table, $E(X) = \sum xp(x) = 2.3$, $E(X^2) = 6.1$, and $V(X) = 6.1 - (2.3)^2 = .81.$ Each lot weighs 5 lbs, so the number of pounds left $= 100 - 5X$. Thus the expected weight left is $E(100 - 5X) = 100 - 5E(X) = 88.5$ lbs, and the variance of the weight left is $V(100 - 5X) = V(-5X) = (-5)^2 V(X) = 25V(X) = 20.25.$

41.   Use the hint: $V(aX + b) = E[((aX + b) - E(aX + b))^2] = \sum[ax + b - E(aX + b)]^2 p(x) =$

$\sum[ax + b - (a\mu + b)]^2 p(x) = \sum[ax - a\mu]^2 p(x) = a^2 \sum(x - \mu)^2 p(x) = a^2 V(X).$

43.   With $a = 1$ and $b = -c$, $E(X - c) = E(aX + b) = a E(X) + b = E(X) - c.$
When $c = \mu$, $E(X - \mu) = E(X) - \mu = \mu - \mu = 0$; i.e., the expected deviation from the mean is zero.

45. $a \leq X \leq b$ means that $a \leq x \leq b$ for all x in the range of X. Hence $ap(x) \leq xp(x) \leq bp(x)$ for all x, and

$$\sum ap(x) \leq \sum xp(x) \leq \sum bp(x)$$
$$a\sum p(x) \leq \sum xp(x) \leq b\sum p(x)$$
$$a \cdot 1 \leq E(X) \leq b \cdot 1$$
$$a \leq E(X) \leq b$$

Section 3.4

47.

   a. $B(4;15,.3) = .515$.

   b. $b(4;15,.3) = B(4;15,.3) - B(3;15,.3) = .219$.

   c. $b(6;15,.7) = B(6;15,.7) - B(5;15,.7) = .012$.

   d. $P(2 \leq X \leq 4) = B(4;15,.3) - B(1;15,.3) = .480$.

   e. $P(2 \leq X) = 1 - P(X \leq 1) = 1 - B(1;15,.3) = .965$.

   f. $P(X \leq 1) = B(1;15,.7) = .000$.

   g. $P(2 < X < 6) = P(2 < X \leq 5) = B(5;15,.3) - B(2;15,.3) = .595$.

49. Let X be the number of "seconds," so $X \sim \text{Bin}(6, .10)$.

   a. $P(X = 1) = \binom{n}{x}p^x(1-p)^{n-x} = \binom{6}{1}(.1)^1(.9)^5 = .3543$.

   b. $P(X \geq 2) = 1 - [P(X = 0) + P(X = 1)] = 1 - \left[\binom{6}{0}(.1)^0(.9)^6 + \binom{6}{1}(.1)^1(.9)^5\right] = 1 - [.5314 + .3543] =$ .1143.

   c. Either 4 or 5 goblets must be selected.

   Select 4 goblets with zero defects: $P(X = 0) = \binom{4}{0}(.1)^0(.9)^4 = .6561$.

   Select 4 goblets, one of which has a defect, and the 5th is good: $\left[\binom{4}{1}(.1)^1(.9)^3\right] \times .9 = .26244$

   So, the desired probability is $.6561 + .26244 = .91854$.

51. Let X be the number of faxes, so $X \sim \text{Bin}(25, .25)$.

   a. $E(X) = np = 25(.25) = 6.25$.

   b. $V(X) = np(1-p) = 25(.25)(.75) = 4.6875$, so $SD(X) = 2.165$.

   c. $P(X > 6.25 + 2(2.165)) = P(X > 10.58) = 1 - P(X \leq 10.58) = 1 - P(X \leq 10) = 1 - B(10;25,.25) = .030$.

53. Let "success" = has at least one citation and define X = number of individuals with at least one citation. Then $X \sim \text{Bin}(n = 15, p = .4)$.

  a. If at least 10 have no citations (failure), then at most 5 have had at least one (success): $P(X \le 5) = B(5;15,.40) = .403$.

  b. Half of 15 is 7.5, so less than half means 7 or fewer: $P(X \le 7) = B(7;15,.40) = .787$.

  c. $P(5 \le X \le 10) = P(X \le 10) - P(X \le 4) = .991 - .217 = .774$.

55. Let "success" correspond to a telephone that is submitted for service while under warranty and must be replaced. Then $p = P(\text{success}) = P(\text{replaced | submitted}) \cdot P(\text{submitted}) = (.40)(.20) = .08$. Thus X, the number among the company's 10 phones that must be replaced, has a binomial distribution with $n = 10$ and $p = .08$, so $P(X = 2) = \binom{10}{2}(.08)^2(.92)^8 = .1478$.

57. Let X = the number of flashlights that work, and let event B = {battery has acceptable voltage}. Then $P(\text{flashlight works}) = P(\text{both batteries work}) = P(B)P(B) = (.9)(.9) = .81$. We have assumed here that the batteries' voltage levels are independent. Finally, $X \sim \text{Bin}(10, .81)$, so $P(X \ge 9) = P(X = 9) + P(X = 10) = .285 + .122 = .407$.

59. In this example, $X \sim \text{Bin}(25, p)$ with p unknown.

  a. $P(\text{rejecting claim when } p = .8) = P(X \le 15 \text{ when } p = .8) = B(15; 25, .8) = .017$.

  b. $P(\underline{\text{not}} \text{ rejecting claim when } p = .7) = P(X > 15 \text{ when } p = .7) = 1 - P(X \le 15 \text{ when } p = .7) =$
  $= 1 - B(15; 25, .7) = 1 - .189 = .811$.
  For $p = .6$, this probability is $= 1 - B(15; 25, .6) = 1 - .575 = .425$.

  c. The probability of rejecting the claim when $p = .8$ becomes $B(14; 25, .8) = .006$, smaller than in **a** above. However, the probabilities of **b** above increase to .902 and .586, respectively. So, by changing 15 to 14, we're making it less likely that we will reject the claim when it's true (p really is $\ge .8$), but more likely that we'll "fail" to reject the claim when it's false (p really is $< .8$).

61. If topic A is chosen, then $n = 2$. When $n = 2$, $P(\text{at least half received}) = P(X \ge 1) = 1 - P(X = 0) =$
$1 - \binom{2}{0}(.9)^0(.1)^2 = .99$.

If topic B is chosen, then $n = 4$. When $n = 4$, $P(\text{at least half received}) = P(X \ge 2) = 1 - P(X \le 1) =$
$1 - \left[\binom{4}{0}(.9)^0(.1)^4 + \binom{4}{1}(.9)^1(.1)^3 \right] = .9963$.
Thus topic B should be chosen if $p = .9$.

However, if $p = .5$, then the probabilities are .75 for A and .6875 for B (using the same method as above), so now A should be chosen.

63.

  a. $b(x; n, 1 - p) = \binom{n}{x}(1 - p)^x(p)^{n-x} = \binom{n}{n-x}(p)^{n-x}(1 - p)^x = b(n-x; n, p)$.

  Conceptually, $P(x \text{ S's when } P(S) = 1 - p) = P(n-x \text{ F's when } P(F) = p)$, since the two events are identical, but the labels S and F are arbitrary and so can be interchanged (if $P(S)$ and $P(F)$ are also interchanged), yielding $P(n-x \text{ S's when } P(S) = 1 - p)$ as desired.

b. Use the conceptual idea from **a**: $B(x; n, 1-p) = P(\underline{\text{at most }} x \text{ S's when } P(S) = 1-p) = P(\underline{\text{at least }} n-x \text{ F's when } P(F) = p)$, since these are the same event
$= P(\underline{\text{at least }} n-x \text{ S's when } P(S) = p)$, since the S and F labels are arbitrary
$= 1 - P(\underline{\text{at most }} n-x-1 \text{ S's when } P(S) = p) = 1 - B(n-x-1; n, p)$.

c. Whenever $p > .5$, $(1-p) < .5$ so probabilities involving X can be calculated using the results **a** and **b** in combination with tables giving probabilities only for $p \le .5$.

65.

a. Although there are three payment methods, we are only concerned with S = uses a debit card and F = does not use a debit card. Thus we can use the binomial distribution. So, if X = the number of customers who use a debit card, $X \sim \text{Bin}(n = 100, p = .2)$. From this, $E(X) = np = 100(.2) = 20$, and $V(X) = npq = 100(.2)(1-.2) = 16$.

b. With S = doesn't pay with cash, $n = 100$ and $p = .7$, so $\mu = np = 100(.7) = 70$, and $V = 21$.

67. When $n = 20$ and $p = .5$, $\mu = 10$ and $\sigma = 2.236$, so $2\sigma = 4.472$ and $3\sigma = 6.708$.
The inequality $|X - 10| \ge 4.472$ is satisfied if either $X \le 5$ or $X \ge 15$, or
$P(|X - \mu| \ge 2\sigma) = P(X \le 5 \text{ or } X \ge 15) = .021 + .021 = .042$. The inequality $|X - 10| \ge 6.708$ is satisfied if either $X \le 3$ or $X \ge 17$, so $P(|X - \mu| \ge 3\sigma) = P(X \le 3 \text{ or } X \ge 17) = .001 + .001 = .002$.

Section 3.5

69. According to the problem description, X is hypergeometric with $n = 6$, $N = 12$, and $M = 7$.

a. $P(X = 5) = \dfrac{\binom{7}{5}\binom{5}{1}}{\binom{12}{6}} = \dfrac{105}{924} = .114$.

b. $P(X \le 4) = 1 - P(X > 4) = 1 - [P(X = 5) + P(X = 6)] = 1 - \left[\dfrac{\binom{7}{5}\binom{5}{1}}{\binom{12}{6}} + \dfrac{\binom{7}{6}\binom{5}{0}}{\binom{12}{6}} \right] =$

$1 - [.114 + .007] = 1 - .121 = .879$.

c. $E(X) = n \cdot \dfrac{M}{N} = 6 \cdot \dfrac{7}{12} = 3.5$; $V(X) = \left(\dfrac{12-6}{12-1} \right) 6 \left(\dfrac{7}{12} \right) \left(1 - \dfrac{7}{12} \right) = 0.795$; $\sigma = 0.892$. So,

$P(X > \mu + \sigma) = P(X > 3.5 + 0.892) = P(X > 4.392) = P(X = 5 \text{ or } 6) = .121$ (from part **b**).

d. We can approximate the hypergeometric distribution with the binomial if the population size and the number of successes are large. Here, $n = 15$ and $M/N = 40/400 = .1$, so $h(x; 15, 40, 400) \approx b(x; 15, .10)$. Using this approximation, $P(X \le 5) \approx B(5; 15, .10) = .998$ from the binomial tables. (This agrees with the exact answer to 3 decimal places.)

71.

a. Possible values of X are 5, 6, 7, 8, 9, 10. (In order to have less than 5 of the granite, there would have to be more than 10 of the basaltic). X is hypergeometric, with $n = 15$, $N = 20$, and $M = 10$. So, the pmf of X is

$$p(x) = h(x;\ 15, 10, 20) = \frac{\binom{10}{x}\binom{10}{15-x}}{\binom{20}{15}}.$$

The pmf is also provided in table form below.

x	5	6	7	8	9	10
$p(x)$	.0163	.1354	.3483	.3483	.1354	.0163

b. P(all 10 of one kind or the other) $= P(X = 5) + P(X = 10) = .0163 + .0163 = .0326$.

c. $\mu = n \cdot \dfrac{M}{N} = 15 \cdot \dfrac{10}{20} = 7.5$; $V(X) = \left(\dfrac{20-15}{20-1}\right)15\left(\dfrac{10}{20}\right)\left(1 - \dfrac{10}{20}\right) = .9868$; $\sigma = .9934$.

$\mu \pm \sigma = 7.5 \pm .9934 = (6.5066, 8.4934)$, so we want $P(6.5066 < X < 8.4934)$. That equals $P(X = 7) + P(X = 8) = .3483 + .3483 = .6966$.

73.

a. The successes here are the top $M = 10$ pairs, and a sample of $n = 10$ pairs is drawn from among the N

$= 20$. The probability is therefore $h(x;\ 10, 10, 20) = \dfrac{\binom{10}{x}\binom{10}{10-x}}{\binom{20}{10}}.$

b. Let $X =$ the number among the top 5 who play east-west. (Now, $M = 5$.)
Then P(all of top 5 play the same direction) $= P(X = 5) + P(X = 0) =$

$$h(5;\ 10, 5, 20) + h(5;\ 10, 5, 20) = \frac{\binom{5}{5}\binom{15}{5}}{\binom{20}{10}} + \frac{\binom{5}{0}\binom{15}{10}}{\binom{20}{10}} = .033\ .$$

c. Generalizing from earlier parts, we now have $N = 2n$; $M = n$. The probability distribution of X is hypergeometric: $p(x) = h(x;\ n, n, 2n) = \dfrac{\binom{n}{x}\binom{n}{n-x}}{\binom{2n}{n}}$ for $x = 0, 1, \ldots, n$. Also,

$$E(X) = n \cdot \frac{n}{2n} = \frac{1}{2}n \text{ and } V(X) = \left(\frac{2n-n}{2n-1}\right) \cdot n \cdot \frac{n}{2n}\left(1 - \frac{n}{2n}\right) = \frac{n^2}{4(2n-1)}.$$

75.

a. With S = a female child and F = a male child, let X = the number of F's before the 2nd S. Then
$$P(X = x) = nb(x; 2, .5) = \binom{x+2-1}{2-1}(.5)^2(1-.5)^x = (x+1)(.5)^{x+2}.$$

b. $P(\text{exactly 4 children}) = P(\text{exactly 2 males} = P(X = 2) = nb(2; 2, .5) = (2+1)(.5)^4 = .188.$

c. $P(\text{at most 4 children}) = P(X \le 2) = \sum_{x=0}^{2} nb(x; 2, .5) = .25 + .25 + .188 = .688.$

d. $E(X) = \dfrac{r(1-p)}{p} = \dfrac{2(1-.5)}{.5} = 2$, so the expected number of children is equal to
$E(X + 2) = E(X) + 2 = 4.$

77. This is identical to an experiment in which a single family has children until exactly 6 females have been born (since $p = .5$ for each of the three families). So,
$$p(x) = nb(x; 6, .5) = \binom{x+5}{5}(.5)^6(1-.5)^x = \binom{x+5}{5}(.5)^{6+x}. \text{ Also, } E(X) = \frac{r(1-p)}{p} = \frac{6(1-.5)}{.5} = 6; \text{ notice this is}$$
just $2 + 2 + 2$, the sum of the expected number of males born to each family.

Section 3.6

79. All these solutions are found using the cumulative Poisson table, $F(x; \mu) = F(x; 5)$.
a. $P(X \le 8) = F(8; 5) = .932.$

b. $P(X = 8) = F(8; 5) - F(7; 5) = .065.$

c. $P(X \ge 9) = 1 - P(X \le 8) = .068.$

d. $P(5 \le X \le 8) = F(8; 5) - F(4; 5) = .492.$

e. $P(5 < X < 8) = F(7; 5) - F(5; 5) = .867 - .616 = .251.$

81. Let $X \sim \text{Poisson}(\mu = 20)$.
a. $P(X \le 10) = F(10; 20) = .011.$

b. $P(X > 20) = 1 - F(20; 20) = 1 - .559 = .441.$

c. $P(10 \le X \le 20) = F(20; 20) - F(9; 20) = .559 - .005 = .554;$
$P(10 < X < 20) = F(19; 20) - F(10; 20) = .470 - .011 = .459.$

d. $E(X) = \mu = 20$, so $\sigma = \sqrt{20} = 4.472$. Therefore, $P(\mu - 2\sigma < X < \mu + 2\sigma) =$
$P(20 - 8.944 < X < 20 + 8.944) \qquad = P(11.056 < X < 28.944) = P(X \le 28) - P(X \le 11) =$
$F(28; 20) - F(11; 20) = .966 - .021 = .945.$

83. The exact distribution of X is binomial with $n = 1000$ and $p = 1/200$; we can approximate this distribution by the Poisson distribution with $\mu = np = 5$.

   a. $P(5 \le X \le 8) = F(8; 5) - F(4; 5) = .492.$

   b. $P(X \ge 8) = 1 - P(X \le 7) = 1 - F(7; 5) = 1 - .867 = .133.$

85.

   a. $\mu = 8$ when $t = 1$, so $P(X = 6) = \dfrac{e^{-8}8^6}{6!} = .122$; $P(X \ge 6) = 1 - F(5; 8) = .809$; and

      $P(X \ge 10) = 1 - F(9; 8) = .283.$

   b. $t = 90$ min $= 1.5$ hours, so $\mu = 12$; thus the expected number of arrivals is 12 and the standard deviation is $\sigma = \sqrt{12} = 3.464.$

   c. $t = 2.5$ hours implies that $\mu = 20$. So, $P(X \ge 20) = 1 - F(19; 20) = .530$ and
      $P(X \le 10) = F(10; 20) = .011.$

87.

   a. For a two hour period the parameter of the distribution is $\mu = \alpha t = (4)(2) = 8$,

      so $P(X = 10) = \dfrac{e^{-8}8^{10}}{10!} = .099.$

   b. For a 30-minute period, $\alpha t = (4)(.5) = 2$, so $P(X = 0) = \dfrac{e^{-2}2^0}{0!} = .135.$

   c. The expected value is simply $E(X) = \alpha t = 2.$

89. In this example, $\alpha =$ rate of occurrence $= 1/($mean time between occurrences$) = 1/.5 = 2.$

   a. For a two-year period, $\mu = \alpha t = (2)(2) = 4$ loads.

   b. Apply a Poisson model with $\mu = 4$: $P(X > 5) = 1 - P(X \le 5) = 1 - F(5; 4) = 1 - .785 = .215.$

   c. For $\alpha = 2$ and the value of t unknown, $P($no loads occur during the period of length $t) =$

      $P(X = 0) = \dfrac{e^{-2t}(2t)^0}{0!} = e^{-2t}$. Solve for t: $e^{-2t} \le .1 \Rightarrow -2t \le \ln(.1) \Rightarrow t \ge 1.1513$ years.

91.

   a. For a quarter-acre (.25 acre) plot, the mean parameter is $\mu = (80)(.25) = 20$, so $P(X \le 16) = F(16; 20) = .221.$

   b. The expected number of trees is $\alpha \cdot ($area$) = 80$ trees/acre $(85,000$ acres$) = 6,800,000$ trees.

   c. The area of the circle is $\pi r^2 = \pi(.1)^2 = .01\pi = .031416$ square miles, which is equivalent to $.031416(640) = 20.106$ acres. Thus X has a Poisson distribution with parameter $\mu = \alpha(20.106) = 80(20.106) = 1608.5$. That is, the pmf of X is the function $p(x; 1608.5)$.

93.

a. No events occur in the time interval $(0, t + \Delta t)$ if and only if no events occur in $(0, t)$ and no events occur in $(t, t + \Delta t)$. Since it's assumed the numbers of events in non-overlapping intervals are independent (Assumption 3),

$P(\text{no events in } (0, t + \Delta t)) = P(\text{no events in } (0, t)) \cdot P(\text{no events in } (t, t + \Delta t)) \Rightarrow$

$P_0(t + \Delta t) = P_0(t) \cdot P(\text{no events in } (t, t + \Delta t)) = P_0(t) \cdot [1 - \alpha\Delta t - o(\Delta t)]$ by Assumption 2.

b. Rewrite **a** as $P_0(t + \Delta t) = P_0(t) - P_0(t)[\alpha\Delta t + o(\Delta t)]$, so $P_0(t + \Delta t) - P_0(t) = -P_0(t)[\alpha\Delta t + o(\Delta t)]$ and

$\dfrac{P_0(t+\Delta t) - P_0(t)}{\Delta t} = -\alpha P_0(t) - P_0(t) \cdot \dfrac{o(\Delta t)}{\Delta t}$. Since $\dfrac{o(\Delta t)}{\Delta t} \to 0$ as $\Delta t \to 0$ and the left-hand side of the

equation converges to $\dfrac{dP_0(t)}{dt}$ as $\Delta t \to 0$, we find that $\dfrac{dP_0(t)}{dt} = -\alpha P_0(t)$.

c. Let $P_0(t) = e^{-\alpha t}$. Then $\dfrac{dP_0(t)}{dt} = \dfrac{d}{dt}[e^{-\alpha t}] = -\alpha e^{-\alpha t} = -\alpha P_0(t)$, as desired. (This suggests that the probability of zero events in $(0, t)$ for a process defined by Assumptions 1-3 is equal to $e^{-\alpha t}$.)

d. Similarly, the product rule implies $\dfrac{d}{dt}\left[\dfrac{e^{-\alpha t}(\alpha t)^k}{k!}\right] = \dfrac{-\alpha e^{-\alpha t}(\alpha t)^k}{k!} + \dfrac{k\alpha e^{-\alpha t}(\alpha t)^{k-1}}{k!} =$

$-\alpha\dfrac{e^{-\alpha t}(\alpha t)^k}{k!} + \alpha\dfrac{e^{-\alpha t}(\alpha t)^{k-1}}{(k-1)!} = -\alpha P_k(t) + \alpha P_{k-1}(t)$, as desired.

Supplementary Exercises

95.

a. We'll find $p(1)$ and $p(4)$ first, since they're easiest, then $p(2)$. We can then find $p(3)$ by subtracting the others from 1.

$p(1) = P(\text{exactly one suit}) = P(\text{all} \spadesuit) + P(\text{all} \heartsuit) + P(\text{all} \diamondsuit) + P(\text{all} \clubsuit) =$

$4 \cdot P(\text{all} \spadesuit) = 4 \cdot \dfrac{\dbinom{13}{5}\dbinom{39}{0}}{\dbinom{52}{5}} = .00198$, since there are 13 $\spadesuit$s and 39 other cards.

$p(4) = 4 \cdot P(2\spadesuit, 1\heartsuit, 1\diamondsuit, 1\clubsuit) = 4 \cdot \dfrac{\dbinom{13}{2}\dbinom{13}{1}\dbinom{13}{1}\dbinom{13}{1}}{\dbinom{52}{5}} = .26375$.

$p(2) = P(\text{all} \heartsuit\text{s and} \spadesuit\text{s, with} \geq \text{one of each}) + \ldots + P(\text{all} \diamondsuit\text{s and} \clubsuit\text{s with} \geq \text{one of each}) =$

$\dbinom{4}{2} \cdot P(\text{all} \heartsuit\text{s and} \spadesuit\text{s, with} \geq \text{one of each}) =$

$6 \cdot [P(1\heartsuit \text{ and } 4\spadesuit) + P(2\heartsuit \text{ and } 3\spadesuit) + P(3\heartsuit \text{ and } 2\spadesuit) + P(4\heartsuit \text{ and } 1\spadesuit)] =$

$6 \cdot \left[2 \cdot \dfrac{\dbinom{13}{4}\dbinom{13}{1}}{\dbinom{52}{5}} + 2 \cdot \dfrac{\dbinom{13}{3}\dbinom{13}{2}}{\dbinom{52}{5}}\right] = 6\left[\dfrac{18,590 + 44,616}{2,598,960}\right] = .14592$.

Finally, $p(3) = 1 - [p(1) + p(2) + p(4)] = .58835$.

54

b. $\mu = \sum\limits_{x=1}^{4} x \cdot p(x) = 3.114$; $\sigma^2 = \left[\sum\limits_{x=1}^{4} x^2 \cdot p(x) \right] - (3.114)^2 = .405 \Rightarrow \sigma = .636.$

97.

    a. From the description, $X \sim \text{Bin}(15, .75)$. So, the pmf of X is $b(x; 15, .75)$.

    b. $P(X > 10) = 1 - P(X \le 10) = 1 - B(10; 15, .75) = 1 - .314 = .686.$

    c. $P(6 \le X \le 10) = B(10; 15, .75) - B(5; 15, .75) = .314 - .001 = .313.$

    d. $\mu = (15)(.75) = 11.75$, $\sigma^2 = (15)(.75)(.25) = 2.81.$

    e. Requests can all be met if and only if $X \le 10$, and $15 - X \le 8$, i.e. iff $7 \le X \le 10$. So,
       $P(\text{all requests met}) = P(7 \le X \le 10) = B(10; 15, .75) - B(6; 15, .75) = .310.$

99.   Let $X =$ the number of components out of 5 that function, so $X \sim \text{Bin}(5, .9)$. Then a 3-out-of 5 system works when X is at least 3, and $P(X \ge 3) = 1 - P(X \le 2) = 1 - B(2; 5, .9) = .991.$

101.

    a. $X \sim \text{Bin}(n = 500, p = .005)$. Since n is large and p is small, X can be approximated by a Poisson distribution with $\mu = np = 2.5$. The approximate pmf of X is $p(x; 2.5) = \dfrac{e^{-2.5} 2.5^x}{x!}$.

    b. $P(X = 5) = \dfrac{e^{-2.5} 2.5^5}{5!} = .0668.$

    c. $P(X \ge 5) = 1 - P(X \le 4) = 1 - p(4; 2.5) = 1 - .8912 = .1088.$

103.   Let Y denote the number of tests carried out.
For $n = 3$, possible Y values are 1 and 4. $P(Y = 1) = P(\text{no one has the disease}) = (.9)^3 = .729$ and $P(Y = 4) = 1 - .729 = .271$, so $E(Y) = (1)(.729) + (4)(.271) = 1.813$, as contrasted with the 3 tests necessary without group testing.
For $n = 5$, possible values of Y are 1 and 6. $P(Y = 1) = P(\text{no one has the disease}) = (.9)^5 = .5905$, so $P(Y = 6) = 1 - .5905 = .4095$ and $E(Y) = (1)(.5905) + (6)(.4095) = 3.0475$, less than the 5 tests necessary without group testing.

105. $p(2) = P(X = 2) = P(SS) = p^2$, and $p(3) = P(FSS) = (1 - p)p^2.$

For $x \ge 4$, consider the first $x - 3$ trials and the last 3 trials separately. To have $X = x$, it must be the case that the last three trials were FSS, and that two-successes-in-a-row was <u>not</u> already seen in the first $x - 3$ tries.

The probability of the first event is simply $(1 - p)p^2$.
The second event occurs if two-in-a-row hadn't occurred after 2 or 3 or ... or $x - 3$ tries. The probability of this second event equals $1 - [p(2) + p(3) + \ldots + p(x - 3)]$. (For $x = 4$, the probability in brackets is empty; for $x = 5$, it's $p(2)$; for $x = 6$, it's $p(2) + p(3)$; and so on.)

Finally, since trials are independent, $P(X = x) = (1 - [p(2) + \ldots + p(x - 3)]) \cdot (1 - p)p^2$.

For $p = .9$, the pmf of X up to $x = 8$ is shown below.

x	2	3	4	5	6	7	8
$p(x)$	.81	.081	.081	.0154	.0088	.0023	.0010

So, $P(X \leq 8) = p(2) + \ldots + p(8) = .9995$.

107.

a. Let event A = seed carries single spikelets, and event B = seed produces ears with single spikelets. Then $P(A \cap B) = P(A) \cdot P(B \mid A) = (.40)(.29) = .116$.
Next, let X = the number of seeds out of the 10 selected that meet the condition $A \cap B$. Then $X \sim$ Bin(10, .116). So, $P(X = 5) = \binom{10}{5}(.116)^5(.884)^5 = .002857$.

b. For any one seed, the event of interest is B = seed produces ears with single spikelets. Using the law of total probability, $P(B) = P(A \cap B) + P(A' \cap B) = (.40)(.29) + (.60)(.26) = .272$.
Next, let Y = the number out of the 10 seeds that meet condition B. Then $Y \sim$ Bin(10, .272). $P(Y = 5) = \binom{10}{5}(.272)^5(1-.272)^5 = .0767$, while

$$P(Y \leq 5) = \sum_{y=0}^{5} \binom{10}{y}(.272)^y(1-.272)^{10-y} = .041813 + \ldots + .076719 = .97024.$$

109.

a. $P(X = 0) = F(0; 2)$ or $\dfrac{e^{-2}2^0}{0!} = 0.135$.

b. Let S = an operator who receives no requests. Then the number of operators that receive no requests follows a Bin($n = 5$, $p = .135$) distribution. So, $P(4\ S\text{'s in 5 trials}) = b(4; 5, .135) = \binom{5}{4}(.135)^4(.865)^1 = .00144$.

c. For any non-negative integer x, $P(\text{all operators receive exactly } x \text{ requests}) =$
$P(\text{first operator receives x}) \cdot \ldots \cdot P(\text{fifth operator receives x}) = [p(x; 2)]^5 = \left[\dfrac{e^{-2}2^x}{x!}\right]^5 = \dfrac{e^{-10}2^{5x}}{(x!)^5}$.
Then, $P(\text{all receive the same number}) = P(\text{all receive 0 requests}) + P(\text{all receive 1 request}) + P(\text{all receive 2 requests}) + \ldots = \sum_{x=0}^{\infty} \dfrac{e^{-10}2^{5x}}{(x!)^5}$.

111. The number of magazine copies sold is X so long as X is no more than five; otherwise, all five copies are sold. So, mathematically, the number sold is min(X, 5), and $E[\min(x, 5)] = \sum_{x=0}^{\infty} \min(x,5)p(x;4) = 0p(0; 4) +$

$1p(1; 4) + 2p(2; 4) + 3p(3; 4) + 4p(4; 4) + \sum_{x=5}^{\infty} 5p(x;4) =$

$1.735 + 5\sum_{x=5}^{\infty} p(x;4) = 1.735 + 5\left[1 - \sum_{x=0}^{4} p(x;4)\right] = 1.735 + 5[1 - F(4; 4)] = 3.59$.

113.

a. No, since the probability of a "success" is not the same for all tests.

b. There are four ways exactly three could have positive results. Let D represent those with the disease and D' represent those without the disease.

Combination		Probability
D	D'	
0	3	$\left[\binom{5}{0}(.2)^0(.8)^5\right] \cdot \left[\binom{5}{3}(.9)^3(.1)^2\right]$
		$=(.32768)(.0729)=.02389$
1	2	$\left[\binom{5}{1}(.2)^1(.8)^4\right] \cdot \left[\binom{5}{2}(.9)^2(.1)^3\right]$
		$=(.4096)(.0081)=.00332$
2	1	$\left[\binom{5}{2}(.2)^2(.8)^3\right] \cdot \left[\binom{5}{1}(.9)^1(.1)^4\right]$
		$=(.2048)(.00045)=.00009216$
3	0	$\left[\binom{5}{3}(.2)^3(.8)^2\right] \cdot \left[\binom{5}{0}(.9)^0(.1)^5\right]$
		$=(.0512)(.00001)=.000000512$

Adding up the probabilities associated with the four combinations yields 0.0273.

115.

a. Notice that $p(x; \mu_1, \mu_2) = .5\, p(x; \mu_1) + .5\, p(x; \mu_2)$, where both terms $p(x; \mu_i)$ are Poisson pmfs. Since both pmfs are ≥ 0, so is $p(x; \mu_1, \mu_2)$. That verifies the first requirement.

Next, $\sum_{x=0}^{\infty} p(x; \mu_1, \mu_2) = .5\sum_{x=0}^{\infty} p(x; \mu_1) + .5\sum_{x=0}^{\infty} p(x; \mu_2) = .5 + .5 = 1$, so the second requirement for a pmf is met. Therefore, $p(x; \mu_1, \mu_2)$ is a valid pmf.

b. $E(X) = \sum_{x=0}^{\infty} x \cdot p(x; \mu_1, \mu_2) = \sum_{x=0}^{\infty} x[.5\,p(x; \mu_1) + .5\,p(x; \mu_2)] = .5\sum_{x=0}^{\infty} x \cdot p(x; \mu_1) + .5\sum_{x=0}^{\infty} x \cdot p(x; \mu_2) = .5E(X_1) +$

$.5E(X_2)$, where $X_i \sim \text{Poisson}(\mu_i)$. Therefore, $E(X) = .5\mu_1 + .5\mu_2$.

c. This requires using the variance shortcut. Using the same method as in **b**,

$E(X^2) = .5\sum_{x=0}^{\infty} x^2 \cdot p(x; \mu_1) + .5\sum_{x=0}^{\infty} x^2 \cdot p(x; \mu_2) = .5E(X_1^2) + .5E(X_2^2)$. For any Poisson rv,

$E(X^2) = V(X) + [E(X)]^2 = \mu + \mu^2$, so $E(X^2) = .5(\mu_1 + \mu_1^2) + .5(\mu_2 + \mu_2^2)$.

Finally, $V(X) = .5(\mu_1 + \mu_1^2) + .5(\mu_2 + \mu_2^2) - [.5\mu_1 + .5\mu_2]^2$, which can be simplified to equal $.5\mu_1 + .5\mu_2 + .25(\mu_1 - \mu_2)^2$.

d. Simply replace the weights .5 and .5 with .6 and .4, so $p(x; \mu_1, \mu_2) = .6\,p(x; \mu_1) + .4\,p(x; \mu_2)$.

117. $P(X=j) = \sum_{i=1}^{10} P(\text{arm on track } i \cap X=j) = \sum_{i=1}^{10} P(X=j \mid \text{arm on } i) \cdot p_i =$

$\sum_{i=1}^{10} P(\text{next seek at } i+j+1 \text{ or } i-j-1) \cdot p_i = \sum_{i=1}^{10} (p_{i+j+1} + p_{i-j-1}) p_i$, where in the summation we take $p_k = 0$ if $k < 0$ or $k > 10$.

119. Using the hint, $\sum_{\text{all } x} (x-\mu)^2 p(x) \geq \sum_{x:|x-\mu| \geq k\sigma} (x-\mu)^2 p(x) \geq \sum_{x:|x-\mu| \geq k\sigma} (k\sigma)^2 p(x) = k^2 \sigma^2 \sum_{x:|x-\mu| \geq k\sigma} p(x)$.

The left-hand side is, by definition, σ^2. On the other hand, the summation on the right-hand side represents $P(|X-\mu| \geq k\sigma)$.

So $\sigma^2 \geq k^2 \sigma^2 \cdot P(|X-\mu| \geq k\sigma)$, whence $P(|X-\mu| \geq k\sigma) \leq 1/k^2$.

121.

a. Let $A_1 = \{\text{voice}\}$, $A_2 = \{\text{data}\}$, and $X =$ duration of a call. Then $E(X) = E(X|A_1)P(A_1) + E(X|A_2)P(A_2) = 3(.75) + 1(.25) = 2.5$ minutes.

b. Let $X =$ the number of chips in a cookie. Then $E(X) = E(X|i = 1)P(i = 1) + E(X| i = 2)P(i = 2) + E(X| i = 3)P(i = 3)$. If X is Poisson, then its mean is the specified μ — that is, $E(X|i) = i + 1$. Therefore, $E(X) = 2(.20) + 3(.50) + 4(.30) = 3.1$ chips.

CHAPTER 4

Section 4.1

1.

    a. The pdf is the straight-line function graphed below on [3, 5]. The function is clearly non-negative; to verify its integral equals 1, compute:

$$\int_3^5 (.075x + .2)\,dx = .0375x^2 + .2x\Big]_3^5 = (.0375(5)^2 + .2(5)) - (.0375(3)^2 + .2(3))$$
$$= 1.9375 - .9375 = 1$$

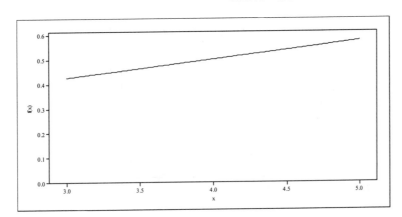

    b. $P(X \le 4) = \int_3^4 (.075x + .2)\,dx = .0375x^2 + .2x\Big]_3^4 = (.0375(4)^2 + .2(4)) - (.0375(3)^2 + .2(3))$

    $= 1.4 - .9375 = .4625$. Since X is a continuous rv, $P(X < 4) = P(X \le 4) = .4625$ as well.

    c. $P(3.5 \le X \le 4.5) = \int_{3.5}^{4.5} (.075x + .2)\,dx = .0375x^2 + .2x\Big]_{3.5}^{4.5} = \cdots = .5$.

    $P(4.5 < X) = P(4.5 \le X) = \int_{4.5}^5 (.075x + .2)\,dx = .0375x^2 + .2x\Big]_{4.5}^5 = \cdots = .278125$.

3.

    a.

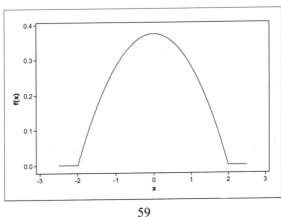

59

b. $P(X > 0) = \int_0^2 .09375(4 - x^2)dx = .09375\left(4x - \frac{x^3}{3}\right)\Big]_0^2 = .5$.

This matches the symmetry of the pdf about $x = 0$.

c. $P(-1 < X < 1) = \int_{-1}^1 .09375(4 - x^2)dx = .6875$.

d. $P(X < -.5 \text{ or } X > .5) = 1 - P(-.5 \le X \le .5) = 1 - \int_{-.5}^{.5} .09375(4 - x^2)dx = 1 - .3672 = .6328$.

5.

a. $1 = \int_{-\infty}^\infty f(x)dx = \int_0^2 kx^2 dx = \frac{kx^3}{3}\Big]_0^2 = \frac{8k}{3} \Rightarrow k = \frac{3}{8}$.

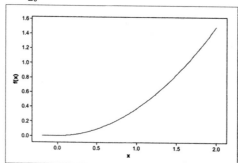

b. $P(0 \le X \le 1) = \int_0^1 \frac{3}{8}x^2 dx = \frac{1}{8}x^3\Big]_0^1 = \frac{1}{8} = .125$.

c. $P(1 \le X \le 1.5) = \int_1^{1.5} \frac{3}{8}x^2 dx = \frac{1}{8}x^3\Big]_1^{1.5} = \frac{1}{8}\left(\frac{3}{2}\right)^3 - \frac{1}{8}(1)^3 = \frac{19}{64} = .296875$.

d. $P(X \ge 1.5) = 1 - \int_{.5}^2 \frac{3}{8}x^2 dx = \frac{1}{8}x^3\Big]_{1.5}^2 = \frac{1}{8}(2)^3 - \frac{1}{8}(1.5)^3 = .578125$.

7.

a. $f(x) = \frac{1}{10}$ for $25 \le x \le 35$ and $= 0$ otherwise.

b. $P(X > 33) = \int_{33}^{35} \frac{1}{10}dx = .2$.

c. $E(X) = \int_{25}^{35} x \cdot \frac{1}{10}dx = \frac{x^2}{20}\Big]_{25}^{35} = 30$. (This is obvious, since the distribution is symmetric and 30 is the midpoint.) 30 ± 2 is from 28 to 32 minutes:
$P(28 < X < 32) = \int_{28}^{32} \frac{1}{10}dx = \frac{1}{10}x\Big]_{28}^{32} = .4$.

d. $P(a \le x \le a + 2) = \int_a^{a+2} \frac{1}{10}dx = .2$, since the interval has length 2.

60

9.

a. $P(X \leq 6) = \int_{.5}^{6} .15e^{-.15(x-.5)}dx = .15\int_{0}^{5.5} e^{-.15u} du$ (after the substitution $u = x - .5$)

$= -e^{-.15u}\big]_{0}^{5.5} = 1 - e^{-.825} \approx .562$

b. $P(X > 6) = 1 - P(X \leq 6) = 1 - .562 = .438$. Since X is continuous, $P(X \geq 6) = P(X > 6) = .438$ as well.

c. $P(5 \leq X \leq 6) = \int_{5}^{6} .15e^{-.15(x-.5)} dx = \int_{4.5}^{6.5} .15e^{-.15u} du = -e^{-.15u}\big]_{4.5}^{5.5} = .071$.

Section 4.2

11.

a. $P(X \leq 1) = F(1) = \dfrac{1^2}{4} = .25$.

b. $P(.5 \leq X \leq 1) = F(1) - F(.5) = \dfrac{1^2}{4} - \dfrac{.5^2}{4} = .1875$.

c. $P(X > 1.5) = 1 - P(X \leq 1.5) = 1 - F(1.5) = 1 - \dfrac{1.5^2}{4} = .4375$.

d. $.5 = F(\tilde{\mu}) = \dfrac{\tilde{\mu}^2}{4} \Rightarrow \tilde{\mu}^2 = 2 \Rightarrow \tilde{\mu} = \sqrt{2} \approx 1.414$.

e. $f(x) = F'(x) = \dfrac{x}{2}$ for $0 \leq x < 2$, and $= 0$ otherwise.

f. $E(X) = \int_{-\infty}^{\infty} x \cdot f(x)dx = \int_{0}^{2} x \cdot \dfrac{x}{2}dx = \dfrac{1}{2}\int_{0}^{2} x^2 dx = \dfrac{x^3}{6}\Big]_{0}^{2} = \dfrac{8}{6} \approx 1.333$.

g. $E(X^2) = \int_{-\infty}^{\infty} x^2 f(x)dx = \int_{0}^{2} x^2 \dfrac{x}{2}dx = \dfrac{1}{2}\int_{0}^{2} x^3 dx = \dfrac{x^4}{8}\Big]_{0}^{2} = 2$, so $V(X) = E(X^2) - [E(X)]^2 =$

$2 - \left(\dfrac{8}{6}\right)^2 = \dfrac{8}{36} \approx .222$, and $\sigma_X = \sqrt{.222} = .471$.

h. From **g**, $E(X^2) = 2$.

13.

a. $1 = \int_1^\infty \frac{k}{x^4} dx = k \int_1^\infty x^{-4} dx = \frac{k}{-3} x^{-3} \Big]_1^\infty = 0 - \left(\frac{k}{-3} \right)(1)^{-3} = \frac{k}{3} \Rightarrow k = 3$.

b. For $x \ge 1$, $F(x) = \int_{-\infty}^x f(y) dy = \int_1^x \frac{3}{y^4} dy = -y^{-3} \Big|_1^x = -x^{-3} + 1 = 1 - \frac{1}{x^3}$. For $x < 1$, $F(x) = 0$ since the

distribution begins at 1. Put together, $F(x) = \begin{cases} 0 & x < 1 \\ 1 - \dfrac{1}{x^3} & 1 \le x \end{cases}$.

c. $P(X > 2) = 1 - F(2) = 1 - \frac{7}{8} = \frac{1}{8}$ or .125;

$P(2 < X < 3) = F(3) - F(2) = \left(1 - \frac{1}{27}\right) - \left(1 - \frac{1}{8}\right) = .963 - .875 = .088$.

d. The mean is $E(X) = \int_1^\infty x \left(\frac{3}{x^4} \right) dx = \int_1^\infty \left(\frac{3}{x^3} \right) dx = -\frac{3}{2} x^{-2} \Big|_1^\infty = 0 + \frac{3}{2} = \frac{3}{2} = 1.5$. Next,

$E(X^2) = \int_1^\infty x^2 \left(\frac{3}{x^4} \right) dx = \int_1^\infty \left(\frac{3}{x^2} \right) dx = -3x^{-1} \Big|_1^\infty = 0 + 3 = 3$, so $V(X) = 3 - (1.5)^2 = .75$. Finally, the

standard deviation of X is $\sigma = \sqrt{.75} = .866$.

e. $P(1.5 - .866 < X < 1.5 + .866) = P(.634 < X < 2.366) = F(2.366) - F(.634) = .9245 - 0 = .9245$.

15.

a. Since X is limited to the interval (0, 1), $F(x) = 0$ for $x \le 0$ and $F(x) = 1$ for $x \ge 1$.
For $0 < x < 1$,
$$F(x) = \int_{-\infty}^x f(y) dy = \int_0^x 90y^8 (1 - y) dy = \int_0^x (90y^8 - 90y^9) dy = 10y^9 - 9y^{10} \Big]_0^x = 10x^9 - 9x^{10}.$$
The graphs of the pdf and cdf of X appear below.

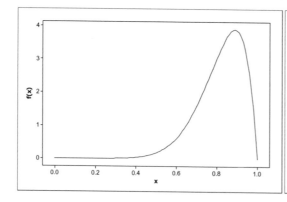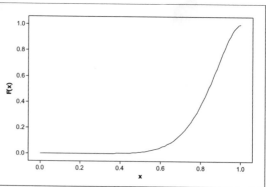

b. $F(.5) = 10(.5)^9 - 9(.5)^{10} = .0107$.

c. $P(.25 < X \le .5) = F(.5) - F(.25) = .0107 - [10(.25)^9 - 9(.25)^{10}] = .0107 - .0000 = .0107$.
Since X is continuous, $P(.25 \le X \le .5) = P(.25 < X \le .5) = .0107$.

d. The 75th percentile is the value of x for which $F(x) = .75$: $10x^9 - 9x^{10} = .75 \Rightarrow x = .9036$ using software.

e. $E(X) = \int_{-\infty}^{\infty} x \cdot f(x)dx = \int_0^1 x \cdot 90x^8(1-x)dx = \int_0^1 (90x^9 - 90x^{10})dx = 9x^{10} - \dfrac{90}{11}x^{11}\Big]_0^1 = 9 - \dfrac{90}{11} = \dfrac{9}{11} = .8182.$

Similarly, $E(X^2) = \int_{-\infty}^{\infty} x^2 \cdot f(x)dx = \int_0^1 x^2 \cdot 90x^8(1-x)dx = \ldots = .6818$, from which $V(X) = .6818 -$ $(.8182)^2 = .0124$ and $\sigma_X = .11134$.

f. $\mu \pm \sigma = (.7068, .9295)$. Thus, $P(\mu - \sigma \le X \le \mu + \sigma) = F(.9295) - F(.7068) = .8465 - .1602 = .6863$, and the probability X is <u>more</u> than 1 standard deviation from its mean value equals $1 - .6863 = 3137$.

17.

a. To find the $(100p)$th percentile, set $F(x) = p$ and solve for x:

$\dfrac{x - A}{B - A} = p \Rightarrow x = A + (B - A)p.$

b. $E(X) = \int_A^B x \cdot \dfrac{1}{B - A}dx = \dfrac{A + B}{2}$, the midpoint of the interval. Also,

$E(X^2) = \dfrac{A^2 + AB + B^2}{3}$, from which $V(X) = E(X^2) - [E(X)]^2 = \ldots = \dfrac{(B - A)^2}{12}$. Finally,

$\sigma_X = \sqrt{V(X)} = \dfrac{B - A}{\sqrt{12}}.$

c. $E(X^n) = \int_A^B x^n \cdot \dfrac{1}{B - A}dx = \dfrac{1}{B - A}\dfrac{x^{n+1}}{n+1}\Big]_A^B = \dfrac{B^{n+1} - A^{n+1}}{(n+1)(B - A)}.$

19.

a. $P(X \le 1) = F(1) = .25[1 + \ln(4)] = .597.$

b. $P(1 \le X \le 3) = F(3) - F(1) = .966 - .597 = .369.$

c. For $x < 0$ or $x > 4$, the pdf is $f(x) = 0$ since X is restricted to $(0, 4)$. For $0 < x < 4$, take the first derivative of the cdf:

$F(x) = \dfrac{x}{4}\left[1 + \ln\left(\dfrac{4}{x}\right)\right] = \dfrac{1}{4}x + \dfrac{\ln(4)}{4}x - \dfrac{1}{4}x\ln(x) \Rightarrow$

$f(x) = F'(x) = \dfrac{1}{4} + \dfrac{\ln(4)}{4} - \dfrac{1}{4}\ln(x) - \dfrac{1}{4}x\dfrac{1}{x} = \dfrac{\ln(4)}{4} - \dfrac{1}{4}\ln(x) = .3466 - .25\ln(x)$

21. $E(\text{area}) = E(\pi R^2) = \int_{-\infty}^{\infty} \pi r^2 f(r)dr = \int_9^{11} \pi r^2 \dfrac{3}{4}\left(1 - (10 - r)^2\right)dr = \cdots = \dfrac{501}{5}\pi = 314.79 \text{ m}^2.$

23. With $X =$ temperature in °C, the temperature in °F equals $1.8X + 32$, so the mean and standard deviation in °F are $1.8\mu_X + 32 = 1.8(120) + 32 = 248$°F and $|1.8|\sigma_X = 1.8(2) = 3.6$°F. Notice that the additive constant, 32, affects the mean but does <u>not</u> affect the standard deviation.

25.

    a. $P(Y \le 1.8\tilde{\mu} + 32) = P(1.8X + 32 \le 1.8\tilde{\mu} + 32) = P(X \le \tilde{\mu}) = .5$ since $\tilde{\mu}$ is the median of X. This shows that $1.8\tilde{\mu} + 32$ is the median of Y.

    b. The 90th percentile for Y equals $1.8\eta(.9) + 32$, where $\eta(.9)$ is the 90th percentile for X. To see this, $P(Y \le 1.8\eta(.9) + 32) = P(1.8X + 32 \le 1.8\eta(.9) + 32) = P(X \le \eta(.9)) = .9$, since $\eta(.9)$ is the 90th percentile of X. This shows that $1.8\eta(.9) + 32$ is the 90th percentile of Y.

    c. When $Y = aX + b$ (i.e. a linear transformation of X) and the $(100p)$th percentile of the X distribution is $\eta(p)$, then the corresponding $(100p)$th percentile of the Y distribution is $a \cdot \eta(p) + b$. This can be demonstrated using the same technique as in **a** and **b** above.

27. Since X is uniform on [0, 360], $E(X) = \dfrac{0 + 360}{2} = 180°$ and $\sigma_X = \dfrac{360 - 0}{\sqrt{12}} = 103.82°$. Using the suggested linear representation of Y, $E(Y) = (2\pi/360)\mu_X - \pi = (2\pi/360)(180) - \pi = 0$ radians, and $\sigma_Y = (2\pi/360)\sigma_X = 1.814$ radians. (In fact, Y is uniform on $[-\pi, \pi]$.)

Section 4.3

29.

    a. .9838 is found in the 2.1 row and the .04 column of the standard normal table so $c = 2.14$.

    b. $P(0 \le Z \le c) = .291 \Rightarrow \Phi(c) - \Phi(0) = .2910 \Rightarrow \Phi(c) - .5 = .2910 \Rightarrow \Phi(c) = .7910 \Rightarrow$ from the standard normal table, $c = .81$.

    c. $P(c \le Z) = .121 \Rightarrow 1 - P(Z < c) = .121 \Rightarrow 1 - \Phi(c) = .121 \Rightarrow \Phi(c) = .879 \Rightarrow c = 1.17$.

    d. $P(-c \le Z \le c) = \Phi(c) - \Phi(-c) = \Phi(c) - (1 - \Phi(c)) = 2\Phi(c) - 1 = .668 \Rightarrow \Phi(c) = .834 \Rightarrow c = 0.97$.

    e. $P(c \le |Z|) = 1 - P(|Z| < c) = 1 - [\Phi(c) - \Phi(-c)] = 1 - [2\Phi(c) - 1] = 2 - 2\Phi(c) = .016 \Rightarrow \Phi(c) = .992 \Rightarrow c = 2.41$.

31. By definition, z_α satisfies $\alpha = P(Z \ge z_\alpha) = 1 - P(Z < z_\alpha) = 1 - \Phi(z_\alpha)$, or $\Phi(z_\alpha) = 1 - \alpha$.

    a. $\Phi(z_{.0055}) = 1 - .0055 = .9945 \Rightarrow z_{.0055} = 2.54$.

    b. $\Phi(z_{.09}) = .91 \Rightarrow z_{.09} \approx 1.34$.

    c. $\Phi(z_{.663}) = .337 \Rightarrow z_{.633} \approx -.42$.

33.

    a. $P(X \le 50) = P\left(Z \le \dfrac{50 - 46.8}{1.75}\right) = P(Z \le 1.83) = \Phi(1.83) = .9664$.

    b. $P(X \ge 48) = P\left(Z \ge \dfrac{48 - 46.8}{1.75}\right) = P(Z \ge 0.69) = 1 - \Phi(0.69) = 1 - .7549 = .2451$.

    c. The mean and standard deviation aren't important here. The probability a normal random variable is within 1.5 standard deviations of its mean equals $P(-1.5 \le Z \le 1.5) = \Phi(1.5) - \Phi(-1.5) = .9332 - .0668 = .8664$.

35.

    a. $P(X \geq 10) = P(Z \geq .43) = 1 - \Phi(.43) = 1 - .6664 = .3336$.
        Since X is continuous, $P(X > 10) = P(X \geq 10) = .3336$.

    b. $P(X > 20) = P(Z > 4) \approx 0$.

    c. $P(5 \leq X \leq 10) = P(-1.36 \leq Z \leq .43) = \Phi(.43) - \Phi(-1.36) = .6664 - .0869 = .5795$.

    d. $P(8.8 - c \leq X \leq 8.8 + c) = .98$, so $8.8 - c$ and $8.8 + c$ are at the 1st and the 99th percentile of the given
        distribution, respectively. The 99th percentile of the standard normal distribution satisfies $\Phi(z) = .99$,
        which corresponds to $z = 2.33$.
        So, $8.8 + c = \mu + 2.33\sigma = 8.8 + 2.33(2.8) \Rightarrow c = 2.33(2.8) = 6.524$.

    e. From **a**, $P(X > 10) = .3336$, so $P(X \leq 10) = 1 - .3336 = .6664$. For four independent selections,
        $P(\text{at least one diameter exceeds } 10) = 1 - P(\text{none of the four exceeds } 10) =$
        $1 - P(\text{first doesn't} \cap \ldots \text{fourth doesn't}) = 1 - (.6664)(.6664)(.6664)(.6664)$ by independence $=$
        $1 - (.6664)^4 = .8028$.

37.

    a. $P(X = 105) = 0$, since the normal distribution is continuous;
        $P(X < 105) = P(Z < 0.2) = P(Z \leq 0.2) = \Phi(0.2) = .5793$;
        $P(X \leq 105) = .5793$ as well, since X is continuous.

    b. No, the answer does not depend on μ or σ. For any normal rv, $P(|X - \mu| > \sigma) = P(|Z| > 1) =$
        $P(Z < -1 \text{ or } Z > 1) = 2P(Z < -1)$ by symmetry $= 2\Phi(-1) = 2(.1587) = .3174$.

    c. From the table, $\Phi(z) = .1\% = .001 \Rightarrow z = -3.09 \Rightarrow x = 104 - 3.09(5) = 88.55$ mmol/L. The smallest
        .1% of chloride concentration values are those less than 88.55 mmol/L

39.

    a. $\mu + \sigma \cdot (91\text{st percentile from standard normal}) = 30 + 5(1.34) = 36.7$.

    b. Similarly, since the 6th percentile of a standard normal distribution is around $z = -1.555$, the 6th
        percentile of this normal distribution is $30 + 5(-1.555) = 22.225$.

    c. We desire the 90th percentile. Since the 90th percentile of a standard normal distribution is around $z =$
        1.28, the 90th percentile of this normal distribution is $30 + 1.28(0.14) = 3.179\mu\text{m}$.

41.   For a single drop, $P(\text{damage}) = P(X < 100) = P\left(Z < \dfrac{100 - 200}{30} \right) = P(Z < -3.33) = .0004$. So, the

      probability of <u>no</u> damage on any single drop is $1 - .0004 = .9996$, and
      $P(\text{at least one among five is damaged}) = 1 - P(\text{none damaged}) = 1 - (.9996)^5 = 1 - .998 = .002$.

43.   Since 1.28 is the 90th z-percentile ($z_{.1} = 1.28$) and -1.645 is the 5th z-percentile ($z_{.05} = 1.645$), the given
      information implies that $\mu + 1.28\sigma = 10.256$ and $\mu - 1.645\sigma = 9.671$.
      Solve: By subtracting the equations, $2.925\sigma = .585$, so $\sigma = .2$, and then $\mu = 10$.

45. With $\mu = .500$ inches, the acceptable range for the diameter is between .496 and .504 inches, so unacceptable bearings will have diameters smaller than .496 or larger than .504.
The new distribution has $\mu = .499$ and $\sigma = .002$.

$$P(X < .496 \text{ or } X > .504) = P\left(Z < \frac{.496 - .499}{.002}\right) + P\left(Z > \frac{.504 - .499}{.002}\right) = P(Z < -1.5) + P(Z > 2.5) =$$

$\Phi(-1.5) + [1 - \Phi(2.5)] = .073$. 7.3% of the bearings will be unacceptable.

47. The stated condition implies that 99% of the area under the normal curve with $\mu = 12$ and $\sigma = 3.5$ is to the left of $c - 1$, so $c - 1$ is the 99th percentile of the distribution. Since the 99th percentile of the standard normal distribution is $z = 2.33$, $c - 1 = \mu + 2.33\sigma = 20.155$, and $c = 21.155$.

49.

    a. $P(X > 4000) = P\left(Z > \frac{4000 - 3432}{482}\right) = P(Z > 1.18) = 1 - \Phi(1.18) = 1 - .8810 = .1190$;

$$P(3000 < X < 4000) = P\left(\frac{3000 - 3432}{482} < Z < \frac{4000 - 3432}{482}\right) = \Phi(1.18) - \Phi(-.90) = .8810 - .1841 = .6969.$$

    b. $P(X < 2000 \text{ or } X > 5000) = P\left(Z < \frac{2000 - 3432}{482}\right) + P\left(Z > \frac{5000 - 3432}{482}\right)$

$$= \Phi(-2.97) + [1 - \Phi(3.25)] = .0015 + .0006 = .0021.$$

    c. We will use the conversion 1 lb = 454 g, then 7 lbs = 3178 grams, and we wish to find

$$P(X > 3178) = P\left(Z > \frac{3178 - 3432}{482}\right) = 1 - \Phi(-.53) = .7019.$$

    d. We need the top .0005 and the bottom .0005 of the distribution. Using the z table, both .9995 and .0005 have multiple z values, so we will use a middle value, ± 3.295. Then $3432 \pm 3.295(482) = 1844$ and 5020. The most extreme .1% of all birth weights are less than 1844 g and more than 5020 g.

    e. Converting to pounds yields a mean of 7.5595 lbs and a standard deviation of 1.0608 lbs. Then

$$P(X > 7) = P\left(Z > \frac{7 - 7.5595}{1.0608}\right) = 1 - \Phi(-.53) = .7019.$$ This yields the same answer as in part **c**.

51. $P(|X - \mu| \geq \sigma) = 1 - P(|X - \mu| < \sigma) = 1 - P(\mu - \sigma < X < \mu + \sigma) = 1 - P(-1 \leq Z \leq 1) = .3174$.
Similarly, $P(|X - \mu| \geq 2\sigma) = 1 - P(-2 \leq Z \leq 2) = .0456$ and $P(|X - \mu| \geq 3\sigma) = .0026$.
These are considerably less than the bounds 1, .25, and .11 given by Chebyshev.

53. $p = .5 \Rightarrow \mu = 12.5$ & $\sigma^2 = 6.25$; $p = .6 \Rightarrow \mu = 15$ & $\sigma^2 = 6$; $p = .8 \Rightarrow \mu = 20$ and $\sigma^2 = 4$. These mean and standard deviation values are used for the normal calculations below.

    a. For the binomial calculation, $P(15 \leq X \leq 20) = B(20; 25, p) - B(14; 25, p)$.

p	$P(15 \leq X \leq 20)$	$P(14.5 \leq \text{Normal} \leq 20.5)$
.5	= .212	= $P(.80 \leq Z \leq 3.20)$ = .2112
.6	= .577	= $P(-.20 \leq Z \leq 2.24)$ = .5668
.8	= .573	= $P(-2.75 \leq Z \leq .25)$ = .5957

66

b. For the binomial calculation, $P(X \leq 15) = B(15; 25, p)$.

p	$P(X \leq 15)$	$P(\text{Normal} \leq 15.5)$
.5	= .885	$= P(Z \leq 1.20) = .8849$
.6	= .575	$= P(Z \leq .20) = .5793$
.8	= .017	$= P(Z \leq -2.25) = .0122$

c. For the binomial calculation, $P(X \geq 20) = 1 - B(19; 25, p)$.

p	$P(X \geq 20)$	$P(\text{Normal} \geq 19.5)$
.5	= .002	$= P(Z \geq 2.80) = .0026$
.6	= .029	$= P(Z \geq 1.84) = .0329$
.8	= .617	$= P(Z \geq -0.25) = .5987$

55. Use the normal approximation to the binomial, with a continuity correction. With $p = .75$ and $n = 500$, $\mu = np = 375$, and $\sigma = 9.68$. So, $\text{Bin}(500, .75) \approx N(375, 9.68)$.

a. $P(360 \leq X \leq 400) = P(359.5 \leq X \leq 400.5) = P(-1.60 \leq Z \leq 2.58) = \Phi(2.58) - \Phi(-1.60) = .9409$.

b. $P(X < 400) = P(X \leq 399.5) = P(Z \leq 2.53) = \Phi(2.53) = .9943$.

57.

a. For any $a > 0$, $F_Y(y) = P(Y \leq y) = P(aX + b \leq y) = P\left(X \leq \frac{y-b}{a}\right) = F_X\left(\frac{y-b}{a}\right)$. This, in turn, implies

$$f_Y(y) = \frac{d}{dy} F_Y(y) = \frac{d}{dy} F_X\left(\frac{y-b}{a}\right) = \frac{1}{a} f_X\left(\frac{y-b}{a}\right).$$

Now let X have a normal distribution. Applying this rule,

$$f_Y(y) = \frac{1}{a}\frac{1}{\sqrt{2\pi}\sigma}\exp\left(-\frac{((y-b)/a - \mu)^2}{2\sigma^2}\right) = \frac{1}{\sqrt{2\pi}a\sigma}\exp\left(-\frac{(y-b-a\mu)^2}{2a^2\sigma^2}\right).$$ This is the pdf of a normal

distribution. In particular, from the exponent we can read that the mean of Y is $E(Y) = a\mu + b$ and the variance of Y is $V(Y) = a^2\sigma^2$. These match the usual rescaling formulas for mean and variance. (The same result holds when $a < 0$.)

b. Temperature in °F would also be normal, with a mean of $1.8(115) + 32 = 239°F$ and a variance of $1.8^2 2^2 = 12.96$ (i.e., a standard deviation of 3.6°F).

Section 4.4

59.

a. $E(X) = \frac{1}{\lambda} = 1$.

b. $\sigma = \frac{1}{\lambda} = 1$.

c. $P(X \leq 4) = 1 - e^{-(1)(4)} = 1 - e^{-4} = .982$.

d. $P(2 \leq X \leq 5) = (1 - e^{-(1)(5)}) - (1 - e^{-(1)(2)}) = e^{-2} - e^{-5} = .129$.

61. Note that a mean value of 2.725 for the exponential distribution implies $\lambda = \dfrac{1}{2.725}$. Let X denote the duration of a rainfall event.

a. $P(X \geq 2) = 1 - P(X < 2) = 1 - P(X \leq 2) = 1 - F(2; \lambda) = 1 - [1 - e^{-(1/2.725)(2)}] = e^{-2/2.725} = .4800$;
$P(X \leq 3) = F(3; \lambda) = 1 - e^{-(1/2.725)(3)} = .6674; P(2 \leq X \leq 3) = .6674 - .4800 = .1874.$

b. For this exponential distribution, $\sigma = \mu = 2.725$, so $P(X > \mu + 2\sigma) =$
$P(X > 2.725 + 2(2.725)) = P(X > 8.175) = 1 - F(8.175; \lambda) = e^{-(1/2.725)(8.175)} = e^{-3} = .0498.$
On the other hand, $P(X < \mu - \sigma) = P(X < 2.725 - 2.725) = P(X < 0) = 0$, since an exponential random variable is non-negative.

63.

a. If a customer's calls are typically short, the first calling plan makes more sense. If a customer's calls are somewhat longer, then the second plan makes more sense, viz. 99¢ is less than 20min(10¢/min) = $2 for the first 20 minutes under the first (flat-rate) plan.

b. $h_1(X) = 10X$, while $h_2(X) = 99$ for $X \leq 20$ and $99 + 10(X - 20)$ for $X > 20$. With $\mu = 1/\lambda$ for the exponential distribution, it's obvious that $E[h_1(X)] = 10E[X] = 10\mu$. On the other hand,
$$E[h_2(X)] = 99 + 10\int_{20}^{\infty} (x-20)\lambda e^{-\lambda x}\,dx = 99 + \frac{10}{\lambda}e^{-20\lambda} = 99 + 10\mu e^{-20/\mu}.$$
When $\mu = 10$, $E[h_1(X)] = 100¢ = \$1.00$ while $E[h_2(X)] = 99 + 100e^{-2} \approx \$1.13$.
When $\mu = 15$, $E[h_1(X)] = 150¢ = \$1.50$ while $E[h_2(X)] = 99 + 150e^{-4/3} \approx \$1.39$.
As predicted, the first plan is better when expected call length is lower, and the second plan is better when expected call length is somewhat higher.

65.

a. $P(X \leq 5) = F(5; 7) = .238.$

b. $P(X < 5) = P(X \leq 5) = .238$ also, since X is continuous.

c. $P(X > 8) = 1 - P(X \leq 8) = 1 - F(8; 7) = .313.$

d. $P(3 \leq X \leq 8) = F(8; 7) - F(3; 7) = .653.$

e. $P(3 < X < 8) = .653$ also, since X is continuous

f. $P(X < 4 \text{ or } X > 6) = 1 - P(4 \leq X \leq 6) = 1 - [F(6; 7) - F(4; 7)] = .713.$

67. Notice that $\mu = 24$ and $\sigma^2 = 144 \Rightarrow \alpha\beta = 24$ and $\alpha\beta^2 = 144 \Rightarrow \beta = \dfrac{144}{24} = 6$ and $\alpha = \dfrac{24}{\beta} = 4.$

a. $P(12 \leq X \leq 24) = F(4; 4) - F(2; 4) = .424.$

b. $P(X \leq 24) = F(4; 4) = .567$, so while the mean is 24, the median is <u>less</u> than 24, since $P(X \leq \tilde{\mu}) = .5$. This is a result of the positive skew of the gamma distribution.

c. We want a value x for which $F\left(\dfrac{x}{\beta}, \alpha\right) = F\left(\dfrac{x}{6}, 4\right) = .99$. In Table A.4, we see $F(10; 4) = .990$. So $x/6 = 10$, and the 99th percentile is $6(10) = 60.$

d. We want a value t for which $P(X > t) = .005$, i.e. $P(X \le t) = .005$. The left-hand side is the cdf of X, so we really want $F\left(\dfrac{t}{6}, 4\right) = .995$. In Table A.4, $F(11; 4) = .995$, so $t/6 = 11$ and $t = 6(11) = 66$. At 66 weeks, only .5% of all transistors would still be operating.

69.

a. $\{X \ge t\} = \{$the lifetime of the system is at least $t\}$. Since the components are connected in series, this equals $\{$all 5 lifetimes are at least $t\} = A_1 \cap A_2 \cap A_3 \cap A_4 \cap A_5$.

b. Since the events A_i are assumed to be independent, $P(X \ge t) = P(A_1 \cap A_2 \cap A_3 \cap A_4 \cap A_5) = P(A_1) \cdot P(A_2) \cdot P(A_3) \cdot P(A_4) \cdot P(A_5)$. Using the exponential cdf, for any i we have $P(A_i) = P(\text{component lifetime is} \ge t) = 1 - F(t) = 1 - [1 - e^{-.01t}] = e^{-.01t}$.
Therefore, $P(X \ge t) = (e^{-.01t}) \cdots (e^{-.01t}) = e^{-.05t}$, and $F_X(t) = P(X \le t) = 1 - e^{-.05t}$.
Taking the derivative, the pdf of X is $f_X(t) = .05e^{-.05t}$ for $t \ge 0$. Thus X also has an exponential distribution, but with parameter $\lambda = .05$.

c. By the same reasoning, $P(X \le t) = 1 - e^{-n\lambda t}$, so X has an exponential distribution with parameter $n\lambda$.

71.

a. $\{X^2 \le y\} = \{-\sqrt{y} \le X \le \sqrt{y}\}$.

b. $F_Y(y) = P(Y \le y) = P(X^2 \le y) = P(-\sqrt{y} \le X \le \sqrt{y}) = \displaystyle\int_{-\sqrt{y}}^{\sqrt{y}} \frac{1}{\sqrt{2\pi}} e^{-z^2/2} dz$. To find the pdf of Y, use the identity (Leibniz's rule):

$$f_Y(y) = \frac{1}{\sqrt{2\pi}} e^{-(\sqrt{y})^2/2} \cdot \frac{d\sqrt{y}}{dy} - \frac{1}{\sqrt{2\pi}} e^{-(-\sqrt{y})^2/2} \cdot \frac{d(-\sqrt{y})}{dy}$$

$$= \frac{1}{\sqrt{2\pi}} e^{-y/2} \cdot \frac{1}{2\sqrt{y}} - \frac{1}{\sqrt{2\pi}} e^{-y/2} \cdot \frac{-1}{2\sqrt{y}} = \frac{1}{\sqrt{2\pi}} y^{-1/2} e^{-y/2}$$

This is valid for $y > 0$. We recognize this as the chi-squared pdf with $\nu = 1$.

Section 4.5

73.

a. $P(X \le 250) = F(250; 2.5, 200) = 1 - e^{-(250/200)^{2.5}} = 1 - e^{-1.75} = .8257$.
$P(X < 250) = P(X \le 250) = .8257$.
$P(X > 300) = 1 - F(300; 2.5, 200) = e^{-(1.5)^{2.5}} = .0636$.

b. $P(100 \le X \le 250) = F(250; 2.5, 200) - F(100; 2.5, 200) = .8257 - .162 = .6637$.

c. The question is asking for the median, $\tilde{\mu}$. Solve $F(\tilde{\mu}) = .5$: $.5 = 1 - e^{-(\tilde{\mu}/200)^{2.5}} \Rightarrow$
$e^{-(\tilde{\mu}/200)^{2.5}} = .5 \Rightarrow (\tilde{\mu}/200)^{2.5} = -\ln(.5) \Rightarrow \tilde{\mu} = 200(-\ln(.5))^{1/2.5} = 172.727$ hours.

75. Using the substitution $y = \left(\dfrac{x}{\beta}\right)^{\alpha} = \dfrac{x^{\alpha}}{\beta^{\alpha}}$. Then $dy = \dfrac{\alpha x^{\alpha-1}}{\beta^{\alpha}}\, dx$, and $\mu = \displaystyle\int_{0}^{\infty} x \cdot \dfrac{\alpha}{\beta^{\alpha}} x^{\alpha-1} e^{-(x/\beta)^{\alpha}}\, dx =$

$\displaystyle\int_{0}^{\infty} (\beta^{\alpha} y)^{1/\alpha} \cdot e^{-y}\, dy = \beta \int_{0}^{\infty} y^{1/\alpha} e^{-y}\, dy = \beta \cdot \Gamma\left(1 + \dfrac{1}{\alpha}\right)$ by definition of the gamma function.

77.

    a. $E(X) = e^{\mu + \sigma^2/2} = e^{4.82} = 123.97$.

        $V(X) = \left(e^{2(4.5)+.8^2}\right) \cdot \left(e^{-.8} - 1\right) = 13{,}776.53 \Rightarrow \sigma = 117.373$.

    b. $P(X \le 100) = \Phi\left(\dfrac{\ln(100) - 4.5}{.8}\right) = \Phi(0.13) = .5517$.

    c. $P(X \ge 200) = 1 - P(X < 200) = 1 - \Phi\left(\dfrac{\ln(200) - 4.5}{.8}\right) = 1 - \Phi(1.00) = 1 - .8413 = .1587$. Since X is continuous,

        $P(X > 200) = .1587$ as well.

79. Notice that μ_X and σ_X are the mean and standard deviation of the lognormal variable X in this example; they are <u>not</u> the parameters μ and σ which usually refer to the mean and standard deviation of $\ln(X)$. We're given $\mu_X = 10{,}281$ and $\sigma_X/\mu_X = .40$, from which $\sigma_X = .40\mu_X = 4112.4$.

    a. To find the mean and standard deviation of $\ln(X)$, set the lognormal mean and variance equal to the appropriate quantities: $10{,}281 = E(X) = e^{\mu + \sigma^2/2}$ and $(4112.4)^2 = V(X) = e^{2\mu + \sigma^2}(e^{\sigma^2} - 1)$. Square the first equation: $(10{,}281)^2 = e^{2\mu + \sigma^2}$. Now divide the variance by this amount:

        $\dfrac{(4112.4)^2}{(10{,}281)^2} = \dfrac{e^{2\mu + \sigma^2}(e^{\sigma^2} - 1)}{e^{2\mu + \sigma^2}} \Rightarrow e^{\sigma^2} - 1 = (.40)^2 = .16 \Rightarrow \sigma = \sqrt{\ln(1.16)} = .38525$

        That's the standard deviation of $\ln(X)$. Use this in the formula for $E(X)$ to solve for μ:

        $10{,}281 = e^{\mu + (.38525)^2/2} = e^{\mu + .07421} \Rightarrow \mu = 9.164$. That's $E(\ln(X))$.

    b. $P(X \le 15{,}000) = P\left(Z \le \dfrac{\ln(15{,}000) - 9.164}{.38525}\right) = P(Z \le 1.17) = \Phi(1.17) = .8790$.

    c. $P(X \ge \mu_X) = P(X \ge 10{,}281) = P\left(Z \ge \dfrac{\ln(10{,}281) - 9.164}{.38525}\right) = P(Z \ge .19) = 1 - \Phi(0.19) = .4247$. Even though the normal distribution is symmetric, the lognormal distribution is <u>not</u> a symmetric distribution. (See the lognormal graphs in the textbook.) So, the mean and the median of X aren't the same and, in particular, the probability X exceeds its own mean doesn't equal .5.

    d. One way to check is to determine whether $P(X < 17{,}000) = .95$; this would mean 17,000 is indeed the 95th percentile. However, we find that $P(X < 17{,}000) = \Phi\left(\dfrac{\ln(17{,}000) - 9.164}{.38525}\right) = \Phi(1.50) = .9332$, so 17,000 is <u>not</u> the 95th percentile of this distribution (it's the 93.32%ile).

81.

a. $E(X) = e^{5+(.01)/2} = e^{5.005} = 149.157$; $V(X) = e^{10+(.01)} \cdot \left(e^{.01} - 1\right) = 223.594$.

b. $P(X > 125) = 1 - P(X \le 125) = = 1 - P\left(Z \le \dfrac{\ln(125) - 5}{.1}\right) = 1 - \Phi(-1.72) = .9573$.

c. $P(110 \le X \le 125) = \Phi(-1.72) - \Phi\left(\dfrac{\ln(110) - 5}{.1}\right) = .0427 - .0013 = .0414$.

d. In Exercise 80, it was shown that $\tilde{\mu} = e^{\mu}$. Here, $\tilde{\mu} = e^5 = 148.41$.

e. From **b**, $P(X > 125) = .9573$. The number of samples whose strengths exceed 125 follows a binomial distribution with $n = 10$ and $p = .9573$, so the expected number is $np = 10(.9573) = 9.573$.

f. It was shown in Exercise 80 that the $100(1 - \alpha)$th percentile is $e^{\mu + \sigma z_\alpha}$. We wish to find the 5th percentile, so $\alpha = .95$ and $z_{.95} = -1.645$. Therefore, the 5th percentile of this lognormal distribution is $e^{5+(.1)(-1.645)} = 125.90$.

83. Since the standard beta distribution lies on $(0, 1)$, the point of symmetry must be ½, so we require that $f\left(\frac{1}{2} - \mu\right) = f\left(\frac{1}{2} + \mu\right)$. Cancelling out the constants, this implies

$\left(\frac{1}{2} - \mu\right)^{\alpha-1} \left(\frac{1}{2} + \mu\right)^{\beta-1} = \left(\frac{1}{2} + \mu\right)^{\alpha-1} \left(\frac{1}{2} - \mu\right)^{\beta-1}$, which (by matching exponents on both sides) in turn implies that $\alpha = \beta$.

Alternatively, symmetry about ½ requires $\mu = \frac{1}{2}$, so $\dfrac{\alpha}{\alpha + \beta} = .5$. Solving for α gives $\alpha = \beta$.

85.

a. Notice from the definition of the standard beta pdf that, since a pdf must integrate to 1,

$$1 = \int_0^1 \frac{\Gamma(\alpha + \beta)}{\Gamma(\alpha)\Gamma(\beta)} x^{\alpha-1}(1-x)^{\beta-1}\, dx \Rightarrow \int_0^1 x^{\alpha-1}(1-x)^{\beta-1}\, dx = \frac{\Gamma(\alpha)\Gamma(\beta)}{\Gamma(\alpha + \beta)}$$

Using this, $E(X) = \int_0^1 x \cdot \dfrac{\Gamma(\alpha + \beta)}{\Gamma(\alpha)\Gamma(\beta)} x^{\alpha-1}(1-x)^{\beta-1}\, dx = \dfrac{\Gamma(\alpha + \beta)}{\Gamma(\alpha)\Gamma(\beta)} \int_0^1 x^{\alpha}(1-x)^{\beta-1}\, dx =$

$\dfrac{\Gamma(\alpha + \beta)}{\Gamma(\alpha)\Gamma(\beta)} \cdot \dfrac{\Gamma(\alpha+1)\Gamma(\beta)}{\Gamma(\alpha+1+\beta)} = \dfrac{\alpha\Gamma(\alpha)}{\Gamma(\alpha)\Gamma(\beta)} \cdot \dfrac{\Gamma(\alpha+\beta)\Gamma(\beta)}{(\alpha+\beta)\Gamma(\alpha+\beta)} = \dfrac{\alpha}{\alpha + \beta}$.

b. Similarly, $E[(1 - X)^m] = \int_0^1 (1-x)^m \cdot \dfrac{\Gamma(\alpha + \beta)}{\Gamma(\alpha)\Gamma(\beta)} x^{\alpha-1}(1-x)^{\beta-1}\, dx =$

$= \dfrac{\Gamma(\alpha + \beta)}{\Gamma(\alpha)\Gamma(\beta)} \int_0^1 x^{\alpha-1}(1-x)^{m+\beta-1}\, dx = \dfrac{\Gamma(\alpha + \beta)}{\Gamma(\alpha)\Gamma(\beta)} \dfrac{\Gamma(\alpha)\Gamma(m+\beta)}{\Gamma(\alpha+m+\beta)} = \dfrac{\Gamma(\alpha + \beta) \cdot \Gamma(m+\beta)}{\Gamma(\alpha+m+\beta)\Gamma(\beta)}$.

If X represents the proportion of a substance consisting of an ingredient, then $1 - X$ represents the proportion not consisting of this ingredient. For $m = 1$ above,

$E(1 - X) = \dfrac{\Gamma(\alpha + \beta) \cdot \Gamma(1+\beta)}{\Gamma(\alpha+1+\beta)\Gamma(\beta)} = \dfrac{\Gamma(\alpha + \beta) \cdot \beta\Gamma(\beta)}{(\alpha+\beta)\Gamma(\alpha+\beta)\Gamma(\beta)} = \dfrac{\beta}{\alpha + \beta}$.

Section 4.6

87. The given probability plot is quite linear, and thus it is quite plausible that the tension distribution is normal.

89. The z percentile values are as follows: $-1.86, -1.32, -1.01, -0.78, -0.58, -0.40, -0.24, -0.08, 0.08, 0.24,$ $0.40, 0.58, 0.78, 1.01, 1.30,$ and 1.86. The accompanying probability plot is reasonably straight, and thus it would be reasonable to use estimating methods that assume a normal population distribution.

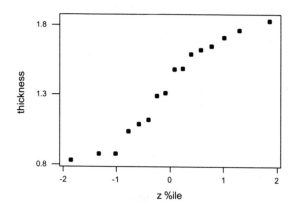

91. The (z percentile, observation) pairs are $(-1.66, .736), (-1.32, .863), (-1.01, .865),$ $(-.78, .913), (-.58, .915), (-.40, .937), (-.24, .983), (-.08, 1.007), (.08, 1.011), (.24, 1.064), (.40, 1.109),$ $(.58, 1.132), (.78, 1.140), (1.01, 1.153), (1.32, 1.253), (1.86, 1.394).$

The accompanying probability plot is straight, suggesting that an assumption of population normality is plausible.

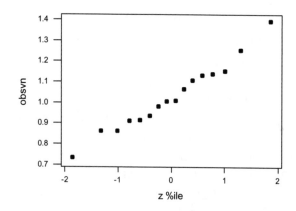

72

93. To check for plausibility of a lognormal population distribution for the rainfall data of Exercise 81 in Chapter 1, take the natural logs and construct a normal probability plot. This plot and a normal probability plot for the original data appear below. Clearly the log transformation gives quite a straight plot, so lognormality is plausible. The curvature in the plot for the original data implies a positively skewed population distribution — like the lognormal distribution.

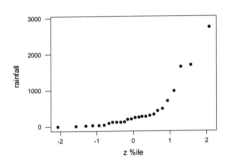 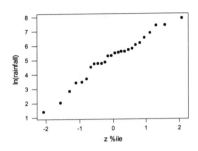

95. The pattern in the plot (below, generated by Minitab) is reasonably linear. By visual inspection alone, it is plausible that strength is normally distributed.

Normal Probability Plot

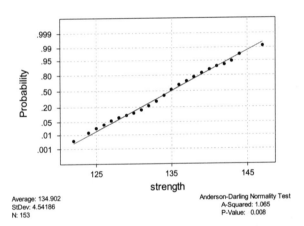

Average: 134.902
StDev: 4.54186
N: 153

Anderson-Darling Normality Test
A-Squared: 1.065
P-Value: 0.008

97. The $(100p)^{th}$ percentile $\eta(p)$ for the exponential distribution with $\lambda = 1$ is given by the formula $\eta(p) = -\ln(1-p)$. With $n = 16$, we need $\eta(p)$ for $p = \frac{0.5}{16}, \frac{1.5}{16}, ..., \frac{15.5}{16}$. These are .032, .398, .170, .247, .330, .421, .521, .633, .758, .901, 1.068, 1.269, 1.520, 1.856, 2.367, 3.466.

The accompanying plot of (percentile, failure time value) pairs exhibits substantial curvature, casting doubt on the assumption of an exponential population distribution.

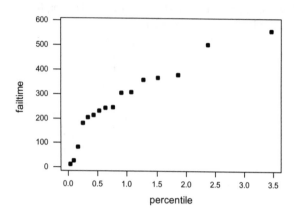

Because λ is a scale parameter (as is σ for the normal family), $\lambda = 1$ can be used to assess the plausibility of the entire exponential family. If we used a different value of λ to find the percentiles, the slope of the graph would change, but not its linearity (or lack thereof).

Supplementary Exercises

99.

a. For $0 \le y \le 25$, $F(y) = \dfrac{1}{24} \int_0^y \left(u - \dfrac{u^2}{12} \right) du = \dfrac{1}{24} \left(\dfrac{u^2}{2} - \dfrac{u^3}{36} \right) \Bigg]_0^y = \dfrac{y^2}{48} - \dfrac{y^3}{864}$. Thus

$$F(y) = \begin{cases} 0 & y < 0 \\ \dfrac{y^2}{48} - \dfrac{y^3}{864} & 0 \le y \le 12 \\ 1 & y > 12 \end{cases}$$

b. $P(Y \le 4) = F(4) = .259$. $P(Y > 6) = 1 - F(6) = .5$.
$P(4 \le X \le 6) = F(6) - F(4) = .5 - .259 = .241$.

c. $E(Y) = \displaystyle\int_0^{12} y \cdot \dfrac{1}{24} y \left(1 - \dfrac{y}{12} \right) dy = \dfrac{1}{24} \int_0^{12} \left(y^2 - \dfrac{y^3}{12} \right) dy = \dfrac{1}{24} \left[\dfrac{y^3}{3} - \dfrac{y^4}{48} \right]_0^{12} = 6$ inches.

$E(Y^2) = \dfrac{1}{24} \displaystyle\int_0^{12} \left(y^3 - \dfrac{y^4}{12} \right) dy = 43.2$, so $V(Y) = 43.2 - 36 = 7.2$.

d. $P(Y < 4 \text{ or } Y > 8) = 1 - P(4 \le Y \le 8) = 1 - [F(8) - F(4)] = .518$.

e. The shorter segment has length equal to $\min(Y, 12 - Y)$, and

$$E[\min(Y, 12 - Y)] = \int_0^{12} \min(y, 12 - y) \cdot f(y) dy = \int_0^6 \min(y, 12 - y) \cdot f(y) dy$$

$$+ \int_6^{12} \min(y, 12 - y) \cdot f(y) dy = \int_0^6 y \cdot f(y) dy + \int_6^{12} (12 - y) \cdot f(y) dy = \dfrac{90}{24} = 3.75 \text{ inches.}$$

101.

a. By differentiation, $f(x) = \begin{cases} x^2 & 0 \le x < 1 \\ \dfrac{7}{4} - \dfrac{3}{4}x & 1 \le x < \dfrac{7}{3} \\ 0 & \text{otherwise} \end{cases}$

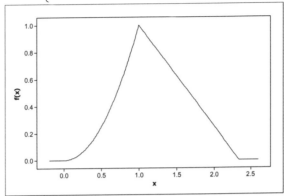

b. $P(.5 \le X \le 2) = F(2) - F(.5) = 1 - \dfrac{1}{2}\left(\dfrac{7}{3} - 2\right)\left(\dfrac{7}{4} - \dfrac{3}{4} \cdot 2\right) - \dfrac{(.5)^3}{3} = \dfrac{11}{12} = .917.$

c. Using the pdf from **a**, $E(X) = \int_0^1 x \cdot x^2 dx + \int_1^{7/3} x \cdot \left(\dfrac{7}{4} - \dfrac{3}{4}x\right) dx = \dfrac{131}{108} = 1.213.$

103.

a. $P(X > 135) = 1 - \Phi\left(\dfrac{135 - 137.2}{1.6}\right) = 1 - \Phi(-1.38) = 1 - .0838 = .9162.$

b. With Y = the number among ten that contain more than 135 oz, $Y \sim \text{Bin}(10, .9162)$.
So, $P(Y \ge 8) = b(8; 10, .9162) + b(9; 10, .9162) + b(10; 10, .9162) = .9549.$

c. We want $P(X > 135) = .95$, i.e. $1 - \Phi\left(\dfrac{135 - 137.2}{\sigma}\right) = .95$ or $\Phi\left(\dfrac{135 - 137.2}{\sigma}\right) = .05$. From the standard

normal table, $\dfrac{135 - 137.2}{\sigma} = -1.65 \Rightarrow \sigma = 1.33.$

105.

a. $P(X > 100) = 1 - \Phi\left(\dfrac{100 - 96}{14}\right) = 1 - \Phi(.29) = 1 - .6141 = .3859.$

b. $P(50 < X < 80) = \Phi\left(\dfrac{80 - 96}{14}\right) - \Phi\left(\dfrac{50 - 96}{14}\right) = \Phi(-1.5) - \Phi(-3.29) = .1271 - .0005 = .1266.$

c. Notice that a and b are the 5th and 95th percentiles, respectively. From the standard normal table, $\Phi(z) = .05 \Rightarrow z = -1.645$, so -1.645 is the 5th percentile of the standard normal distribution. By symmetry, the 95th percentile is $z = 1.645$. So, the desired percentiles of this distribution are $a = 96 + (-1.645)(14) = 72.97$ and $b = 96 + (1.645)(14) = 119.03$. The interval (72.97, 119.03) contains the central 90% of all grain sizes.

107.

a.

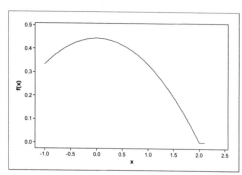

b. $F(x) = 0$ for $x < -1$, and $F(x) = 1$ for $x > 2$. For $-1 \leq x \leq 2$, $F(x) = \int_{-1}^{x} \frac{1}{9}(4 - y^2)\,dy = \frac{11 + 12x - x^3}{27}$. This is graphed below.

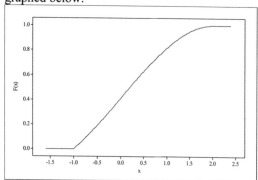

c. The median is 0 iff $F(0) = .5$. Since $F(0) = \frac{11}{27}$, this is not the case. Because $\frac{11}{27} < .5$, the median must be greater than 0. (Looking at the pdf in **a** it's clear that the line $x = 0$ does not evenly divide the distribution, and that such a line must lie to the right of $x = 0$.)

d. Y is a binomial rv, with $n = 10$ and $p = P(X > 1) = 1 - F(1) = \frac{5}{27}$.

109. Below, $\exp(u)$ is alternative notation for e^u.

a. $P(X \leq 150) = \exp\left[-\exp\left(\frac{-(150-150)}{90}\right)\right] = \exp[-\exp(0)] = \exp(-1) = .368$,

$P(X \leq 300) = \exp[-\exp(-1.6667)] = .828$, and $P(150 \leq X \leq 300) = .828 - .368 = .460$.

b. The desired value c is the 90th percentile, so c satisfies

$.9 = \exp\left[-\exp\left(\frac{-(c-150)}{90}\right)\right]$. Taking the natural log of each side twice in succession yields $\frac{-(c-150)}{90}$

$= \ln[-\ln(.9)] = -2.250367$, so $c = 90(2.250367) + 150 = 352.53$.

c. Use the chain rule: $f(x) = F'(x) = \exp\left[-\exp\left(\frac{-(x-\alpha)}{\beta}\right)\right] \cdot -\exp\left(\frac{-(x-\alpha)}{\beta}\right) \cdot -\frac{1}{\beta} =$

$\frac{1}{\beta}\exp\left[-\exp\left(\frac{-(x-\alpha)}{\beta}\right) - \frac{(x-\alpha)}{\beta}\right]$.

d. We wish the value of x for which $f(x)$ is a maximum; from calculus, this is the same as the value of x for which $\ln[f(x)]$ is a maximum, and $\ln[f(x)] = -\ln\beta - e^{-(x-\alpha)/\beta} - \dfrac{(x-\alpha)}{\beta}$. The derivative of $\ln[f(x)]$ is

$$\frac{d}{dx}\left[-\ln\beta - e^{-(x-\alpha)/\beta} - \frac{(x-\alpha)}{\beta}\right] = 0 + \frac{1}{\beta}e^{-(x-\alpha)/\beta} - \frac{1}{\beta}$$; set this equal to 0 and we get $e^{-(x-\alpha)/\beta} = 1$, so

$\dfrac{-(x-\alpha)}{\beta} = 0$, which implies that $x = \alpha$. Thus the mode is α.

e. $E(X) = .5772\beta + \alpha = 201.95$, whereas the mode is $\alpha = 150$ and the median is the solution to $F(x) = .5$.
From **b**, this equals $-90\ln[-\ln(.5)] + 150 = 182.99$.
Since mode < median < mean, the distribution is positively skewed. A plot of the pdf appears below.

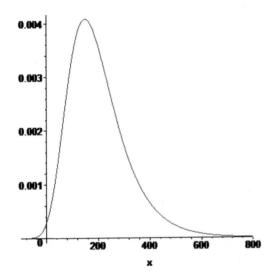

111.

a. From a graph of the normal pdf or by differentiation of the pdf, $x^* = \mu$.

b. No; the density function has constant height for $A \le x \le B$.

c. $f(x; \lambda)$ is largest for $x = 0$ (the derivative at 0 does not exist since f is not continuous there), so $x^* = 0$.

d. $\ln[f(x; \alpha, \beta)] = -\ln(\beta^\alpha) - \ln(\Gamma(\alpha)) + (\alpha-1)\ln(x) - \dfrac{x}{\beta}$, and $\dfrac{d}{dx}\ln[f(x; \alpha, \beta)] = \dfrac{\alpha-1}{x} - \dfrac{1}{\beta}$. Setting this equal to 0 gives the mode: $x^* = (\alpha - 1)\beta$.

e. The chi-squared distribution is the gamma distribution with $\alpha = \nu/2$ and $\beta = 2$. From **d**,
$$x^* = \left(\frac{\nu}{2} - 1\right)(2) = \nu - 2.$$

113.

a. Clearly $f(x; \lambda_1, \lambda_2, p) \geq 0$ for all x. The integral of this function is given by
$$\int_0^\infty \left[p\lambda_1 e^{-\lambda_1 x} + (1-p)\lambda_2 e^{-\lambda_2 x} \right] dx = p\int_0^\infty \lambda_1 e^{-\lambda_1 x} dx + (1-p)\int_0^\infty \lambda_2 e^{-\lambda_2 x} dx = p + (1-p) = 1.$$ (Each of the two integrals represents the integral of an exponential pdf, therefore each of the two integrals integrates to 1.)

b. For $x > 0$, $F(x; \lambda_1, \lambda_2, p) = \int_0^x f(y; \lambda_1, \lambda_2, p) dy = \int_0^x \left[p\lambda_1 e^{-\lambda_1 y} + (1-p)\lambda_2 e^{-\lambda_2 y} \right] dy =$
$$p\int_0^x \lambda_1 e^{-\lambda_1 y} dy + (1-p)\int_0^x \lambda_2 e^{-\lambda_2 y} dy = p(1 - e^{-\lambda_1 x}) + (1-p)(1 - e^{-\lambda_2 x}).$$ For $x \leq 0$, $F(x) = 0$.

c. $E(X) = \int_0^\infty x \cdot \left[p\lambda_1 e^{-\lambda_1 x} + (1-p)\lambda_2 e^{-\lambda_2 x} \right] dx = p\int_0^\infty x\lambda_1 e^{-\lambda_1 x} dx + (1-p)\int_0^\infty x\lambda_2 e^{-\lambda_2 x} dx = \dfrac{p}{\lambda_1} + \dfrac{(1-p)}{\lambda_2}$. (Each of the two integrals represents the expected value of an exponential random variable, which is the reciprocal of λ.)

d. Similarly, $E(X^2) = \dfrac{2p}{\lambda_1^2} + \dfrac{2(1-p)}{\lambda_2^2}$, so $V(X) = \dfrac{2p}{\lambda_1^2} + \dfrac{2(1-p)}{\lambda_2^2} - \left[\dfrac{p}{\lambda_1} + \dfrac{(1-p)}{\lambda_2} \right]^2$.

e. For an exponential rv, $CV = \dfrac{\sigma}{\mu} = \dfrac{1/\lambda}{1/\lambda} = 1$. For X hyperexponential,
$$CV = \frac{\sigma}{\mu} = \frac{\sqrt{E(X^2) - \mu^2}}{\mu} = \sqrt{\frac{E(X^2)}{\mu^2} - 1} = \sqrt{\frac{2p/\lambda_1^2 + 2(1-p)/\lambda_2^2}{\left[p/\lambda_1 + (1-p)/\lambda_2 \right]^2} - 1} =$$
$$\sqrt{\frac{2\left(p\lambda_2^2 + (1-p)\lambda_1^2 \right)}{\left(p\lambda_2 + (1-p)\lambda_1 \right)^2} - 1} = \sqrt{2r - 1}, \text{ where } r = \frac{p\lambda_2^2 + (1-p)\lambda_1^2}{\left(p\lambda_2 + (1-p)\lambda_1 \right)^2}.$$ But straightforward algebra shows that $r > 1$ when $\lambda_1 \neq \lambda_2$, so that $CV > 1$.

f. For the Erlang distribution, $\mu = \dfrac{n}{\lambda}$ and $\sigma = \dfrac{\sqrt{n}}{\lambda}$, so $CV = \dfrac{1}{\sqrt{n}} < 1$ for $n > 1$.

115.

a. Since $\ln\left(\dfrac{I_o}{I_i}\right)$ has a normal distribution, by definition $\dfrac{I_o}{I_i}$ has a lognormal distribution.

b. $P(I_o > 2I_i) = P\left(\dfrac{I_o}{I_i} > 2\right) = P\left(\ln\left(\dfrac{I_o}{I_i}\right) > \ln 2\right) = 1 - P\left(\ln\left(\dfrac{I_o}{I_i}\right) \leq \ln 2\right) = 1 - P(X \leq \ln 2) =$
$1 - \Phi\left(\dfrac{\ln 2 - 1}{.05}\right) = 1 - \Phi(-6.14) = 1$.

c. $E\left(\dfrac{I_o}{I_i}\right) = e^{1 + .0025/2} = 2.72$ and $V\left(\dfrac{I_o}{I_i}\right) = e^{2 + .0025} \cdot \left(e^{.0025} - 1\right) = .0185$.

117. $F(y) = P(Y \le y) = P(\sigma Z + \mu \le y) = P\left(Z \le \dfrac{y-\mu}{\sigma}\right) = \int_{-\infty}^{\frac{y-\mu}{\sigma}} \dfrac{1}{\sqrt{2\pi}} e^{-\frac{1}{2}z^2} dz \Rightarrow$ by the fundamental theorem of

calculus, $f(y) = F'(y) = \dfrac{1}{\sqrt{2\pi}} e^{-\frac{1}{2}\left(\frac{y-\mu}{\sigma}\right)^2} \cdot \dfrac{1}{\sigma} = \dfrac{1}{\sqrt{2\pi}\sigma} e^{-\frac{1}{2}\left(\frac{y-\mu}{\sigma}\right)^2}$, a normal pdf with parameters μ and σ.

119.

    a. $Y = -\ln(X) \Rightarrow x = e^{-y} = k(y)$, so $k'(y) = -e^{-y}$. Thus since $f(x) = 1$, $g(y) = 1 \cdot |-e^{-y}| = e^{-y}$ for $0 < y < \infty$. Y has an exponential distribution with parameter $\lambda = 1$.

    b. $y = \sigma Z + \mu \Rightarrow z = k(y) = \dfrac{y-\mu}{\sigma}$ and $k'(y) = \dfrac{1}{\sigma}$, from which the result follows easily.

    c. $y = h(x) = cx \Rightarrow x = k(y) = \dfrac{y}{c}$ and $k'(y) = \dfrac{1}{c}$, from which the result follows easily.

121.

    a. Assuming the three birthdays are independent and that all 365 days of the calendar year are equally likely, $P(\text{all 3 births occur on March 11}) = \left(\dfrac{1}{365}\right)^3$.

    b. $P(\text{all 3 births on the same day}) = P(\text{all 3 on Jan. 1}) + P(\text{all 3 on Jan. 2}) + \ldots = \left(\dfrac{1}{365}\right)^3 + \left(\dfrac{1}{365}\right)^3 + \ldots = $

    $365\left(\dfrac{1}{365}\right)^3 = \left(\dfrac{1}{365}\right)^2$.

    c. Let X = deviation from due date, so $X \sim N(0, 19.88)$. The baby due on March 15 was 4 days early, and
    $P(X = -4) \approx P(-4.5 < X < -3.5) = \Phi\left(\dfrac{-3.5}{19.88}\right) - \Phi\left(\dfrac{-4.5}{19.88}\right) = $
    $\Phi(-.18) - \Phi(-.237) = .4286 - .4090 = .0196$.
    Similarly, the baby due on April 1 was 21 days early, and $P(X = -21) \approx$
    $\Phi\left(\dfrac{-20.5}{19.88}\right) - \Phi\left(\dfrac{-21.5}{19.88}\right) = \Phi(-1.03) - \Phi(-1.08) = .1515 - .1401 = .0114$.
    Finally, the baby due on April 4 was 24 days early, and $P(X = -24) \approx .0097$.

    Again assuming independence, $P(\text{all 3 births occurred on March 11}) = (.0196)(.0114)(.0097) = .0002145$.

    d. To calculate the probability of the three births happening on any day, we could make similar calculations as in part **c** for each possible day, and then add the probabilities.

123.

a. $F(x) = P(X \leq x) = P\left(-\dfrac{1}{\lambda}\ln(1-U) \leq x\right) = P\left(\ln(1-U) \geq -\lambda x\right) = P\left(1 - U \geq e^{-\lambda x}\right)$

$= P\left(U \leq 1 - e^{-\lambda x}\right) = 1 - e^{-\lambda x}$ since the cdf of a uniform rv on [0, 1] is simply $F(u) = u$. Thus X has an exponential distribution with parameter λ.

b. By taking successive random numbers $u_1, u_2, u_3, \ldots$ and computing $x_i = -\dfrac{1}{10}\ln(1 - u_i)$ for each one, we obtain a sequence of values generated from an exponential distribution with parameter $\lambda = 10$.

125. If $g(x)$ is convex in a neighborhood of μ, then $g(\mu) + g'(\mu)(x - \mu) \leq g(x)$. Replace x by X:
$E[g(\mu) + g'(\mu)(X - \mu)] \leq E[g(X)] \Rightarrow E[g(X)] \geq g(\mu) + g'(\mu)E[(X - \mu)] = g(\mu) + g'(\mu) \cdot 0 = g(\mu)$.
That is, if $g(x)$ is convex, $g(E(X)) \leq E[g(X)]$.

127.

a. $E(X) = 150 + (850 - 150)\dfrac{8}{8+2} = 710$ and $V(X) = \dfrac{(850-150)^2(8)(2)}{(8+2)^2(8+2+1)} = 7127.27 \Rightarrow SD(X) \approx 84.423$.

Using software, $P(|X - 710| \leq 84.423) = P(625.577 \leq X \leq 794.423) =$

$\displaystyle\int_{625.577}^{794.423} \dfrac{1}{700}\dfrac{\Gamma(10)}{\Gamma(8)\Gamma(2)}\left(\dfrac{x-150}{700}\right)^7\left(\dfrac{850-x}{700}\right)^1 dx = .684$.

b. $P(X > 750) = \displaystyle\int_{750}^{850} \dfrac{1}{700}\dfrac{\Gamma(10)}{\Gamma(8)\Gamma(2)}\left(\dfrac{x-150}{700}\right)^7\left(\dfrac{850-x}{700}\right)^1 dx = .376$. Again, the computation of the requested integral requires a calculator or computer.

CHAPTER 5

Section 5.1

1.

 a. $P(X = 1, Y = 1) = p(1,1) = .20.$

 b. $P(X \le 1 \text{ and } Y \le 1) = p(0,0) + p(0,1) + p(1,0) + p(1,1) = .42.$

 c. At least one hose is in use at both islands. $P(X \ne 0 \text{ and } Y \ne 0) = p(1,1) + p(1,2) + p(2,1) + p(2,2) = .70.$

 d. By summing row probabilities, $p_X(x) = .16, .34, .50$ for $x = 0, 1, 2$, By summing column probabilities, $p_Y(y) = .24, .38, .38$ for $y = 0, 1, 2$. $P(X \le 1) = p_X(0) + p_X(1) = .50.$

 e. $p(0,0) = .10$, but $p_X(0) \cdot p_Y(0) = (.16)(.24) = .0384 \ne .10$, so X and Y are not independent.

3.

 a. $p(1,1) = .15$, the entry in the 1st row and 1st column of the joint probability table.

 b. $P(X_1 = X_2) = p(0,0) + p(1,1) + p(2,2) + p(3,3) = .08 + .15 + .10 + .07 = .40.$

 c. $A = \{X_1 \ge 2 + X_2 \cup X_2 \ge 2 + X_1\}$, so $P(A) = p(2,0) + p(3,0) + p(4,0) + p(3,1) + p(4,1) + p(4,2) + p(0,2) + p(0,3) + p(1,3) = .22.$

 d. $P(X_1 + X_2 = 4) = p(1,3) + p(2,2) + p(3,1) + p(4,0) = .17.$
  $P(X_1 + X_2 \ge 4) = P(X_1 + X_2 = 4) + p(4,1) + p(4,2) + p(4,3) + p(3,2) + p(3,3) + p(2,3) = .46.$

5.

 a. $p(3, 3) = P(X = 3, Y = 3) = P(3 \text{ customers, each with 1 package})$
  $= P(\text{ each has 1 package} \mid 3 \text{ customers}) \cdot P(3 \text{ customers}) = (.6)^3 \cdot (.25) = .054.$

 b. $p(4, 11) = P(X = 4, Y = 11) = P(\text{total of 11 packages} \mid 4 \text{ customers}) \cdot P(4 \text{ customers}).$
  Given that there are 4 customers, there are four different ways to have a total of 11 packages: 3, 3, 3, 2 or 3, 3, 2, 3 or 3, 2, 3, 3 or 2, 3, 3, 3. Each way has probability $(.1)^3(.3)$, so $p(4, 11) = 4(.1)^3(.3)(.15) = .00018.$

7.

 a. $p(1,1) = .030.$

 b. $P(X \le 1 \text{ and } Y \le 1) = p(0,0) + p(0,1) + p(1,0) + p(1,1) = .120.$

 c. $P(X = 1) = p(1,0) + p(1,1) + p(1,2) = .100; P(Y = 1) = p(0,1) + \ldots + p(5,1) = .300.$

 d. $P(\text{overflow}) = P(X + 3Y > 5) = 1 - P(X + 3Y \le 5) = 1 - P((X,Y)=(0,0) \text{ or } \ldots \text{or } (5,0) \text{ or } (0,1) \text{ or } (1,1) \text{ or } (2,1)) = 1 - .620 = .380.$

81

e. The marginal probabilities for X (row sums from the joint probability table) are $p_X(0) = .05$, $p_X(1) = .10$, $p_X(2) = .25$, $p_X(3) = .30$, $p_X(4) = .20$, $p_X(5) = .10$; those for Y (column sums) are $p_Y(0) = .5$, $p_Y(1) = .3$, $p_Y(2) = .2$. It is now easily verified that for every (x,y), $p(x,y) = p_X(x) \cdot p_Y(y)$, so X and Y are independent.

9.

a. $1 = \int_{-\infty}^{\infty} \int_{-\infty}^{\infty} f(x,y)\,dx\,dy = \int_{20}^{30} \int_{20}^{30} K(x^2 + y^2)\,dx\,dy = K \int_{20}^{30} \int_{20}^{30} x^2\,dy\,dx + K \int_{20}^{30} \int_{20}^{30} y^2\,dx\,dy$

$= 10K \int_{20}^{30} x^2\,dx + 10K \int_{20}^{30} y^2\,dy = 20K \cdot \left(\dfrac{19,000}{3} \right) \Rightarrow K = \dfrac{3}{380,000}$.

b. $P(X < 26 \text{ and } Y < 26) = \int_{20}^{26} \int_{20}^{26} K(x^2 + y^2)\,dx\,dy = K \int_{20}^{26} \left[x^2 y + \dfrac{y^3}{3} \right]_{20}^{26} dx = K \int_{20}^{26} (6x^2 + 3192)\,dx =$

$K(38,304) = .3024$.

c. The region of integration is labeled *III* below.

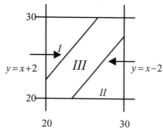

$P(|X - Y| \le 2) = \iint_{III} f(x,y)\,dx\,dy = 1 - \iint_{I} f(x,y)\,dx\,dy - \iint_{II} f(x,y)\,dx\,dy =$

$1 - \int_{20}^{28} \int_{x+2}^{30} f(x,y)\,dy\,dx - \int_{22}^{30} \int_{20}^{x-2} f(x,y)\,dy\,dx = .3593 \text{(after much algebra)}.$

d. $f_X(x) = \int_{-\infty}^{\infty} f(x,y)\,dy = \int_{20}^{30} K(x^2 + y^2)\,dy = 10Kx^2 + K \dfrac{y^3}{3} \Big|_{20}^{30} = 10Kx^2 + .05$, for $20 \le x \le 30$.

e. $f_Y(y)$ can be obtained by substituting y for x in (d); clearly $f(x,y) \ne f_X(x) \cdot f_Y(y)$, so X and Y are not independent.

11.

a. Since X and Y are independent, $p(x,y) = p_X(x) \cdot p_Y(y) = \dfrac{e^{-\mu_1} \mu_1^x}{x!} \cdot \dfrac{e^{-\mu_2} \mu_2^y}{y!} = \dfrac{e^{-\mu_1 - \mu_2} \mu_1^x \mu_2^y}{x! y!}$

for $x = 0, 1, 2, \ldots; y = 0, 1, 2, \ldots$.

b. $P(X + Y \le 1) = p(0,0) + p(0,1) + p(1,0) = \ldots = e^{-\mu_1 - \mu_2} [1 + \mu_1 + \mu_2]$.

c. $P(X + Y = m) = \sum_{k=0}^{m} P(X = k, Y = m - k) = e^{-\mu_1 - \mu_2} \sum_{k=0}^{m} \dfrac{\mu_1^k}{k!} \cdot \dfrac{\mu_2^{m-k}}{(m-k)!} = \dfrac{e^{-\mu_1 - \mu_2}}{m!} \sum_{k=0}^{m} \dfrac{m!}{k!(m-k)!} \mu_1^k \mu_2^{m-k} =$

$\dfrac{e^{-\mu_1 - \mu_2}}{m!} \sum_{k=0}^{m} \binom{m}{k} \mu_1^k \mu_2^{m-k} = \dfrac{e^{-\mu_1 - \mu_2}}{m!} (\mu_1 + \mu_2)^m$ by the binomial theorem. We recognize this as the pmf of a

Poisson random variable with parameter $\mu_1 + \mu_2$. Therefore, the total number of errors, $X + Y$, also has a Poisson distribution, with parameter $\mu_1 + \mu_2$.

13.

a. $f(x,y) = f_X(x) \cdot f_Y(y) = \begin{cases} e^{-x-y} & x \geq 0, y \geq 0 \\ 0 & \text{otherwise} \end{cases}$

b. By independence, $P(X \leq 1 \text{ and } Y \leq 1) = P(X \leq 1) \cdot P(Y \leq 1) = (1 - e^{-1})(1 - e^{-1}) = .400$.

c. $P(X + Y \leq 2) = \int_0^2 \int_0^{2-x} e^{-x-y} dy dx = \int_0^2 e^{-x}\left[1 - e^{-(2-x)}\right] dx = \int_0^2 (e^{-x} - e^{-2}) dx = 1 - e^{-2} - 2e^{-2} = .594$.

d. $P(X + Y \leq 1) = \int_0^1 e^{-x}\left[1 - e^{-(1-x)}\right] dx = 1 - 2e^{-1} = .264$,

so $P(1 \leq X + Y \leq 2) = P(X + Y \leq 2) - P(X + Y \leq 1) = .594 - .264 = .330$.

15.

a. Each X_i has cdf $F(x) = P(X_i \leq x) = 1 - e^{-\lambda x}$. Using this, the cdf of Y is
$F(y) = P(Y \leq y) = P(X_1 \leq y \cup [X_2 \leq y \cap X_3 \leq y])$
$= P(X_1 \leq y) + P(X_2 \leq y \cap X_3 \leq y) - P(X_1 \leq y \cap [X_2 \leq y \cap X_3 \leq y])$
$= (1 - e^{-\lambda y}) + (1 - e^{-\lambda y})^2 - (1 - e^{-\lambda y})^3$ for $y > 0$.

The pdf of Y is $f(y) = F'(y) = \lambda e^{-\lambda y} + 2(1 - e^{-\lambda y})(\lambda e^{-\lambda y}) - 3(1 - e^{-\lambda y})^2(\lambda e^{-\lambda y}) = 4\lambda e^{-2\lambda y} - 3\lambda e^{-3\lambda y}$
for $y > 0$.

b. $E(Y) = \int_0^\infty y \cdot \left(4\lambda e^{-2\lambda y} - 3\lambda e^{-3\lambda y}\right) dy = 2\left(\frac{1}{2\lambda}\right) - \frac{1}{3\lambda} = \frac{2}{3\lambda}$.

17.

a. Let A denote the disk of radius $R/2$. Then $P((X,Y) \text{ lies in } A) = \iint_A f(x,y) dx dy$

$= \iint_A \frac{1}{\pi R^2} dx dy = \frac{1}{\pi R^2} \iint_A dx dy = \frac{\text{area of } A}{\pi R^2} = \frac{\pi (R/2)^2}{\pi R^2} = \frac{1}{4} = .25$. Notice that, since the joint pdf of X and Y is a constant (i.e., (X,Y) is <u>uniform</u> over the disk), it will be the case for any subset A that $P((X,Y)$ lies in $A) = \frac{\text{area of } A}{\pi R^2}$.

b. By the same ratio-of-areas idea, $P\left(-\frac{R}{2} \leq X \leq \frac{R}{2}, -\frac{R}{2} \leq Y \leq \frac{R}{2}\right) = \frac{R^2}{\pi R^2} = \frac{1}{\pi}$. This region is the square depicted in the graph below.

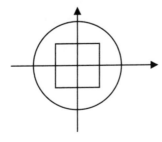

83

c. Similarly, $P\left(-\dfrac{R}{\sqrt{2}} \le X \le \dfrac{R}{\sqrt{2}}, -\dfrac{R}{\sqrt{2}} \le Y \le \dfrac{R}{\sqrt{2}}\right) = \dfrac{2R^2}{\pi R^2} = \dfrac{2}{\pi}$. This region is the slightly larger square depicted in the graph below, whose corners actually touch the circle.

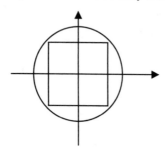

d. $f_X(x) = \displaystyle\int_{-\infty}^{\infty} f(x,y)dy = \int_{-\sqrt{R^2-x^2}}^{\sqrt{R^2-x^2}} \dfrac{1}{\pi R^2}\,dy = \dfrac{2\sqrt{R^2-x^2}}{\pi R^2}$ for $-R \le x \le R$.

Similarly, $f_Y(y) = \dfrac{2\sqrt{R^2-y^2}}{\pi R^2}$ for $-R \le y \le R$. X and Y are <u>not</u> independent, since the joint pdf is not

the product of the marginal pdfs: $\dfrac{1}{\pi R^2} \ne \dfrac{2\sqrt{R^2-x^2}}{\pi R^2} \cdot \dfrac{2\sqrt{R^2-y^2}}{\pi R^2}$.

19. Throughout these solutions, $K = \dfrac{3}{380,000}$, as calculated in Exercise 9.

a. $f_{Y|X}(y\,|\,x) = \dfrac{f(x,y)}{f_X(x)} = \dfrac{K(x^2+y^2)}{10Kx^2+.05}$ for $20 \le y \le 30$.

$f_{X|Y}(x\,|\,y) = \dfrac{f(x,y)}{f_Y(y)} = \dfrac{K(x^2+y^2)}{10Ky^2+.05}$ for $20 \le x \le 30$.

b. $P(Y \ge 25 \,|\, X = 22) = \displaystyle\int_{25}^{30} f_{Y|X}(y\,|\,22)dy = \int_{25}^{30} \dfrac{K((22)^2+y^2)}{10K(22)^2+.05}\,dy = .5559$.

$P(Y \ge 25) = \displaystyle\int_{25}^{30} f_Y(y)dy = \int_{25}^{30}(10Ky^2+.05)dy = .75$. So, given that the right tire pressure is 22 psi, it's much less likely that the left tire pressure is at least 25 psi.

c. $E(Y\,|\,X = 22) = \displaystyle\int_{-\infty}^{\infty} y \cdot f_{Y|X}(y\,|\,22)dy = \int_{20}^{30} y \cdot \dfrac{K((22)^2+y^2)}{10K(22)^2+.05}\,dy = 25.373$ psi.

$E(Y^2\,|\,X = 22) = \displaystyle\int_{20}^{30} y^2 \cdot \dfrac{k((22)^2+y^2)}{10k(22)^2+.05}\,dy = 652.03 \Rightarrow$

$V(Y\,|\,X = 22) = E(Y^2\,|\,X = 22) - [E(Y\,|\,X = 22)]^2 = 652.03 - (25.373)^2 = 8.24 \Rightarrow$
$SD(Y\,|\,X = 22) = 2.87$ psi.

21.

a. $f_{X_3 | X_1, X_2}\left(x_3 \mid x_1, x_2\right) = \dfrac{f(x_1, x_2, x_3)}{f_{X_1, X_2}(x_1, x_2)}$, where $f_{X_1, X_2}(x_1, x_2) = $ the marginal joint pdf of X_1 and X_2, i.e.

$f_{X_1, X_2}(x_1, x_2) = \displaystyle\int_{-\infty}^{\infty} f(x_1, x_2, x_3)\, dx_3$.

b. $f_{X_2, X_3 | X_1}\left(x_2, x_3 \mid x_1\right) = \dfrac{f(x_1, x_2, x_3)}{f_{X_1}(x_1)}$, where $f_{X_1}(x_1) = \displaystyle\int_{-\infty}^{\infty}\int_{-\infty}^{\infty} f(x_1, x_2, x_3)\, dx_2 dx_3$, the marginal pdf of X_1.

Section 5.2

23. $E(X_1 - X_2) = \displaystyle\sum_{x_1=0}^{4}\sum_{x_2=0}^{3}(x_1 - x_2) \cdot p(x_1, x_2) = (0-0)(.08) + (0-1)(.07) + \ldots + (4-3)(.06) = .15.$

Note: It can be shown that $E(X_1 - X_2)$ always equals $E(X_1) - E(X_2)$, so in this case we could also work out the means of X_1 and X_2 from their marginal distributions: $E(X_1) = 1.70$ and $E(X_2) = 1.55$, so $E(X_1 - X_2) = E(X_1) - E(X_2) = 1.70 - 1.55 = .15.$

25. The expected value of X, being uniform on $[L - A, L + A]$, is simply the midpoint of the interval, L. Since Y has the same distribution, $E(Y) = L$ as well. Finally, since X and Y are independent,
$E(\text{area}) = E(XY) = E(X) \cdot E(Y) = L \cdot L = L^2.$

27. The amount of time Annie waits for Alvie, if Annie arrives first, is $Y - X$; similarly, the time Alvie waits for Annie is $X - Y$. Either way, the amount of time the first person waits for the second person is $h(X, Y) = |X - Y|$. Since X and Y are independent, their joint pdf is given by $f_X(x) \cdot f_Y(y) = (3x^2)(2y) = 6x^2 y$. From these, the expected waiting time is

$E[h(X,Y)] = \displaystyle\int_0^1\int_0^1 |x - y| \cdot f(x, y)\,dxdy = \int_0^1\int_0^1 |x - y| \cdot 6x^2 y\,dxdy$

$= \displaystyle\int_0^1\int_0^x (x - y) \cdot 6x^2 y\,dydx + \int_0^1\int_x^1 (x - y) \cdot 6x^2 y\,dydx = \dfrac{1}{6} + \dfrac{1}{12} = \dfrac{1}{4}$ hour, or 15 minutes.

29. $\text{Cov}(X,Y) = -\dfrac{2}{75}$ and $\mu_X = \mu_Y = \dfrac{2}{5}$.

$E(X^2) = \displaystyle\int_0^1 x^2 \cdot f_X(x)dx = 12\int_0^1 x^3(1 - x^2 dx) = \dfrac{12}{60} = \dfrac{1}{5}$, so $V(X) = \dfrac{1}{5} - \left(\dfrac{2}{5}\right)^2 = \dfrac{1}{25}$.

Similarly, $V(Y) = \dfrac{1}{25}$, so $\rho_{X,Y} = \dfrac{-\frac{2}{75}}{\sqrt{\frac{1}{25}} \cdot \sqrt{\frac{1}{25}}} = -\dfrac{50}{75} = -\dfrac{2}{3}$.

31.

a. $E(X) = \displaystyle\int_{20}^{30} x f_X(x)dx = \int_{20}^{30} x\left[10Kx^2 + .05\right]dx = \dfrac{1925}{76} = 25.329 = E(Y)$,

$E(XY) = \displaystyle\int_{20}^{30}\int_{20}^{30} xy \cdot K(x^2 + y^2)dxdy = \dfrac{24375}{38} = 641.447 \Rightarrow$
$\text{Cov}(X, Y) = 641.447 - (25.329)^2 = -.1082.$

b. $E(X^2) = \displaystyle\int_{20}^{30} x^2\left[10Kx^2 + .05\right]dx = \dfrac{37040}{57} = 649.8246 = E(Y^2) \Rightarrow$

$V(X) = V(Y) = 649.8246 - (25.329)^2 = 8.2664 \Rightarrow \rho = \dfrac{-.1082}{\sqrt{(8.2664)(8.2664)}} = -.0131.$

85

33. Since $E(XY) = E(X) \cdot E(Y)$, $\text{Cov}(X, Y) = E(XY) - E(X) \cdot E(Y) = E(X) \cdot E(Y) - E(X) \cdot E(Y) = 0$, and since $\text{Corr}(X, Y) = \dfrac{\text{Cov}(X,Y)}{\sigma_X \sigma_Y}$, then $\text{Corr}(X, Y) = 0$.

35.

a. $\text{Cov}(aX + b, cY + d) = E[(aX + b)(cY + d)] - E(aX + b) \cdot E(cY + d)$
$= E[acXY + adX + bcY + bd] - (aE(X) + b)(cE(Y) + d)$
$= acE(XY) + adE(X) + bcE(Y) + bd - [acE(X)E(Y) + adE(X) + bcE(Y) + bd]$
$= acE(XY) - acE(X)E(Y) = ac[E(XY) - E(X)E(Y)] = ac\text{Cov}(X, Y)$.

b. $\text{Corr}(aX + b, cY + d) = \dfrac{\text{Cov}(aX + b, cY + d)}{SD(aX + b)SD(cY + d)} = \dfrac{ac\text{Cov}(X,Y)}{|a| \cdot |c| SD(X)SD(Y)} = \dfrac{ac}{|ac|}\text{Corr}(X,Y)$. When a and c have the same signs, $ac = |ac|$, and we have $\text{Corr}(aX + b, cY + d) = \text{Corr}(X, Y)$

c. When a and c differ in sign, $|ac| = -ac$, and we have $\text{Corr}(aX + b, cY + d) = -\text{Corr}(X, Y)$.

Section 5.3

37. The joint pmf of X_1 and X_2 is presented below. Each joint probability is calculated using the independence of X_1 and X_2; e.g., $p(25, 25) = P(X_1 = 25) \cdot P(X_2 = 25) = (.2)(.2) = .04$.

		x_1			
$p(x_1, x_2)$		25	40	65	
	25	.04	.10	.06	.2
x_2	40	.10	.25	.15	.5
	65	.06	.15	.09	.3
		.2	.5	.3	

a. For each coordinate in the table above, calculate $\bar{x}$. The six possible resulting $\bar{x}$ values and their corresponding probabilities appear in the accompanying pmf table.

$\bar{x}$	25	32.5	40	45	52.5	65
$p(\bar{x})$	.04	.20	.25	.12	.30	.09

From the table, $E(\bar{X}) = (25)(.04) + 32.5(.20) + ... + 65(.09) = 44.5$. From the original pmf, $\mu = 25(.2) + 40(.5) + 65(.3) = 44.5$. So, $E(\bar{X}) = \mu$.

b. For each coordinate in the joint pmf table above, calculate $s^2 = \dfrac{1}{2-1}\sum_{i=1}^{2}(x_i - \bar{x})^2$. The four possible resulting s^2 values and their corresponding probabilities appear in the accompanying pmf table.

s^2	0	112.5	312.5	800
$p(s^2)$	.38	.20	.30	.12

From the table, $E(S^2) = 0(.38) + ... + 800(.12) = 212.25$. From the original pmf, $\sigma^2 = (25 - 44.5)^2(.2) + (40 - 44.5)^2(.5) + (65 - 44.5)^2(.3) = 212.25$. So, $E(S^2) = \sigma^2$.

39. X is a binomial random variable with $p = .8$. The values of X, then X/n, along with the corresponding probabilities $p(x; 10, .8)$ are displayed in the accompanying pmf table.

x	0	1	2	3	4	5	6	7	8	9	10
x/n	0	.1	.2	.3	.4	.5	.6	.7	.8	.9	1
$p(x/n)$	.000	.000	.000	.001	.005	.027	.088	.201	.302	.269	.107

41.   The tables below delineate all 16 possible (x_1, x_2) pairs, their probabilities, the value of $\bar{x}$ for that pair, and the value of r for that pair. Probabilities are calculated using the independence of X_1 and X_2.

(x_1, x_2)	1,1	1,2	1,3	1,4	2,1	2,2	2,3	2,4
probability	.16	.12	.08	.04	.12	.09	.06	.03
$\bar{x}$	1	1.5	2	2.5	1.5	2	2.5	3
r	0	1	2	3	1	0	1	2

(x_1, x_2)	3,1	3,2	3,3	3,4	4,1	4,2	4,3	4,4
probability	.08	.06	.04	.02	.04	.03	.02	.01
$\bar{x}$	2	2.5	3	3.5	2.5	3	3.5	4
r	2	1	0	1	3	2	1	0

a.   Collecting the $\bar{x}$ values from the table above yields the pmf table below.

$\bar{x}$	1	1.5	2	2.5	3	3.5	4
$p(\bar{x})$	.16	.24	.25	.20	.10	.04	.01

b. $P(\bar{X} \le 2.5) = .16 + .24 + .25 + .20 = .85$.

c.   Collecting the r values from the table above yields the pmf table below.

r	0	1	2	3
$p(r)$	.30	.40	.22	.08

d.   With $n = 4$, there are numerous ways to get a sample average of at most 1.5, since $\bar{X} \le 1.5$ iff the sum of the X_i is at most 6. Listing out all options, $P(\bar{X} \le 1.5) = P(1,1,1,1) + P(2,1,1,1) + \dots + P(1,1,1,2) + P(1,1,2,2) + \dots + P(2,2,1,1) + P(3,1,1,1) + \dots + P(1,1,1,3)$
$= (.4)^4 + 4(.4)^3(.3) + 6(.4)^2(.3)^2 + 4(.4)^2(.2)^2 = .2400$.

43.   The statistic of interest is the fourth spread, or the difference between the medians of the upper and lower halves of the data. The population distribution is uniform with $A = 8$ and $B = 10$. Use a computer to generate samples of sizes $n = 5$, 10, 20, and 30 from a uniform distribution with $A = 8$ and $B = 10$. Keep the number of replications the same (say 500, for example). For each replication, compute the upper and lower fourth, then compute the difference. Plot the sampling distributions on separate histograms for $n = 5$, 10, 20, and 30.

45. Using Minitab to generate the necessary sampling distribution, we can see that as n increases, the distribution slowly moves toward normality. However, even the sampling distribution for $n = 50$ is not yet approximately normal.

$n = 10$

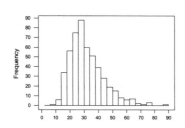

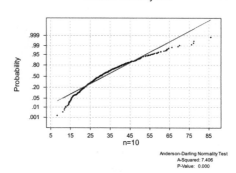

$n = 50$

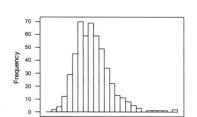

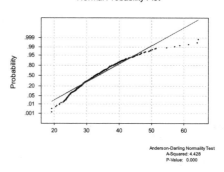

Section 5.4

47.

a. In the previous exercise, we found $E(\overline{X}) = 12$ and $SD(\overline{X}) = .01$ when $n = 16$. If the diameter distribution is normal, then $\overline{X}$ is also normal, so

$$P(\ 11.99 \le \overline{X} \le 12.01) = P\left(\frac{11.99-12}{.01} \le Z \le \frac{12.01-12}{.01}\right) = P(-1 \le Z \le 1)$$

$$= \Phi(1) - \Phi(-1) = .8413 - .1587 = .6826.$$

b. With $n = 25$, $E(\overline{X}) = 12$ but $SD(\overline{X}) = \dfrac{.04}{\sqrt{25}} = .008$. So, $P(\overline{X} > 12.01) = P\left(Z > \dfrac{12.01-12}{.008}\right) =$

$P(Z > 1.25) = 1 - \Phi(1.25) = 1 - .8944 = .1056.$

49.

a. 11 P.M. – 6:50 P.M. = 250 minutes. With $T_o = X_1 + \ldots + X_{40} =$ total grading time,

$\mu_{T_o} = n\mu = (40)(6) = 240$ and $\sigma_{T_o} = \sigma \cdot \sqrt{n} = 37.95$, so $P(T_o \le 250) \approx$

$$P\left(Z \le \frac{250-240}{37.95}\right) = P(Z \le .26) = .6026.$$

b. The sports report begins 260 minutes after he begins grading papers.

$$P(T_0 > 260) = P\left(Z > \frac{260 - 240}{37.95}\right) = P(Z > .53) = .2981.$$

51. Individual times are given by $X \sim N(10, 2)$. For day 1, $n = 5$, and so

$$P(\bar{X} \le 11) = P\left(Z \le \frac{11 - 10}{2/\sqrt{5}}\right) = P(Z \le 1.12) = .8686 .$$

For day 2, $n = 6$, and so

$$P(\bar{X} \le 11) = P(\bar{X} \le 11) = P\left(Z \le \frac{11 - 10}{2/\sqrt{6}}\right) = P(Z \le 1.22) = .8888 .$$

Finally, assuming the results of the two days are independent (which seems reasonable), the probability the sample average is at most 11 min on both days is $(.8686)(.8888) = .7720$.

53.

a. With the values provided,

$$P(\bar{X} \ge 51) = P\left(Z \ge \frac{51 - 50}{1.2/\sqrt{9}}\right) = P(Z \ge 2.5) = 1 - .9938 = .0062 .$$

b. Replace $n = 9$ by $n = 40$, and

$$P(\bar{X} \ge 51) = P\left(Z \ge \frac{51 - 50}{1.2/\sqrt{40}}\right) = P(Z \ge 5.27) \approx 0 .$$

55.

a. With $Y = $ # of tickets, Y has approximately a normal distribution with $\mu = 50$ and $\sigma = \sqrt{\mu} = 7.071$. So, using a continuity correction from [35, 70] to [34.5, 70.5],

$$P(35 \le Y \le 70) \approx P\left(\frac{34.5 - 50}{7.071} \le Z \le \frac{70.5 - 50}{7.071}\right) = P(-2.19 \le Z \le 2.90) = .9838.$$

b. Now $\mu = 5(50) = 250$, so $\sigma = \sqrt{250} = 15.811$.

Using a continuity correction from [225, 275] to [224.5, 275.5], $P(225 \le Y \le 275) \approx$

$$P\left(\frac{224.5 - 250}{15.811} \le Z \le \frac{275.5 - 250}{15.811}\right) = P(-1.61 \le Z \le 1.61) = .8926.$$

57. With the parameters provided, $E(X) = \alpha\beta = 100$ and $V(X) = \alpha\beta^2 = 200$. Using a normal approximation,

$$P(X \le 125) \approx P\left(Z \le \frac{125 - 100}{\sqrt{200}}\right) = P(Z \le 1.77) = .9616.$$

Section 5.5

59.

a. $E(X_1 + X_2 + X_3) = 180$, $V(X_1 + X_2 + X_3) = 45$, $SD(X_1 + X_2 + X_3) = \sqrt{45} = 6.708$.

$P(X_1 + X_2 + X_3 \leq 200) = P\left(Z \leq \dfrac{200 - 180}{6.708}\right) = P(Z \leq 2.98) = .9986$.

$P(150 \leq X_1 + X_2 + X_3 \leq 200) = P(-4.47 \leq Z \leq 2.98) \approx .9986$.

b. $\mu_{\bar{X}} = \mu = 60$ and $\sigma_{\bar{X}} = \dfrac{\sigma_X}{\sqrt{n}} = \dfrac{\sqrt{15}}{\sqrt{3}} = 2.236$, so

$P(\bar{X} \geq 55) = P\left(Z \geq \dfrac{55 - 60}{2.236}\right) = P(Z \geq -2.236) = .9875$ and

$P(58 \leq \bar{X} \leq 62) = P\left(-.89 \leq Z \leq .89\right) = .6266$.

c. $E(X_1 - .5X_2 - .5X_3) = \mu - .5\mu - .5\mu = 0$, while
$V(X_1 - .5X_2 - .5X_3) = \sigma_1^2 + .25\sigma_2^2 + .25\sigma_3^2 = 22.5 \Rightarrow SD(X_1 - .5X_2 - .5X_3) = 4.7434$. Thus,

$P(-10 \leq X_1 - .5X_2 - .5X_3 \leq 5) = P\left(\dfrac{-10 - 0}{4.7434} \leq Z \leq \dfrac{5 - 0}{4.7434}\right) = P(-2.11 \leq Z \leq 1.05) = .8531 - .0174 = $ .8357.

d. $E(X_1 + X_2 + X_3) = 150$, $V(X_1 + X_2 + X_3) = 36 \Rightarrow SD(X_1 + X_2 + X_3) = 6$, so

$P(X_1 + X_2 + X_3 \leq 200) = P\left(Z \leq \dfrac{160 - 150}{6}\right) = P(Z \leq 1.67) = .9525$.

Next, we want $P(X_1 + X_2 \geq 2X_3)$, or, written another way, $P(X_1 + X_2 - 2X_3 \geq 0)$.
$E(X_1 + X_2 - 2X_3) = 40 + 50 - 2(60) = -30$ and $V(X_1 + X_2 - 2X_3) = \sigma_1^2 + \sigma_2^2 + 4\sigma_3^2 = 78 \Rightarrow$
$SD(X_1 + X_2 - 2X_3) = 8.832$, so

$P(X_1 + X_2 - 2X_3 \geq 0) = P\left(Z \geq \dfrac{0 - (-30)}{8.832}\right) = P(Z \geq 3.40) = .0003$.

61.

a. The marginal pmfs of X and Y are given in the solution to Exercise 7, from which $E(X) = 2.8$, $E(Y) = .7$, $V(X) = 1.66$, and $V(Y) = .61$. Thus, $E(X + Y) = E(X) + E(Y) = 3.5$, $V(X + Y) = V(X) + V(Y) = 2.27$, and the standard deviation of $X + Y$ is 1.51.

b. $E(3X + 10Y) = 3E(X) + 10E(Y) = 15.4$, $V(3X + 10Y) = 9V(X) + 100V(Y) = 75.94$, and the standard deviation of revenue is 8.71.

63.

a. $E(X_1) = 1.70$, $E(X_2) = 1.55$, $E(X_1 X_2) = \sum\sum\limits_{x_1\ x_2} x_1 x_2 p(x_1, x_2) = \cdots = 3.33$, so

$\text{Cov}(X_1, X_2) = E(X_1 X_2) - E(X_1)\,E(X_2) = 3.33 - 2.635 = .695$.

b. $V(X_1 + X_2) = V(X_1) + V(X_2) + 2\text{Cov}(X_1, X_2) = 1.59 + 1.0875 + 2(.695) = 4.0675$. This is much larger than $V(X_1) + V(X_2)$, since the two variables are positively correlated.

65.

a. $E(\overline{X} - \overline{Y}) = 0$; $V(\overline{X} - \overline{Y}) = \dfrac{\sigma^2}{25} + \dfrac{\sigma^2}{25} = .0032 \Rightarrow \sigma_{\overline{X} - \overline{Y}} = \sqrt{.0032} = .0566$

$\Rightarrow P(-.1 \le \overline{X} - \overline{Y} \le .1) = P(-1.77 \le Z \le 1.77) = .9232$.

b. $V(\overline{X} - \overline{Y}) = \dfrac{\sigma^2}{36} + \dfrac{\sigma^2}{36} = .0022222 \Rightarrow \sigma_{\overline{X} - \overline{Y}} = .0471$

$\Rightarrow P(-.1 \le \overline{X} - \overline{Y} \le .1) \approx P(-2.12 \le Z \le 2.12) = .9660$. The normal curve calculations are still justified here, even though the populations are not normal, by the Central Limit Theorem (36 is a sufficiently "large" sample size).

67. Letting X_1, X_2, and X_3 denote the lengths of the three pieces, the total length is $X_1 + X_2 - X_3$. This has a normal distribution with mean value $20 + 15 - 1 = 34$ and variance $.25 + .16 + .01 = .42$ from which the standard deviation is $.6481$. Standardizing gives $P(34.5 \le X_1 + X_2 - X_3 \le 35) = P(.77 \le Z \le 1.54) = .1588$.

69.

a. $E(X_1 + X_2 + X_3) = 800 + 1000 + 600 = 2400$.

b. Assuming independence of X_1, X_2, X_3, $V(X_1 + X_2 + X_3) = (16)^2 + (25)^2 + (18)^2 = 1205$.

c. $E(X_1 + X_2 + X_3) = 2400$ as before, but now $V(X_1 + X_2 + X_3)$ $= V(X_1) + V(X_2) + V(X_3) + 2\text{Cov}(X_1, X_2) + 2\text{Cov}(X_1, X_3) + 2\text{Cov}(X_2, X_3) = 1745$, from which the standard deviation is 41.77.

71.

a. $M = a_1 X_1 + a_2 X_2 + W \displaystyle\int_0^{12} x\,dx = a_1 X_1 + a_2 X_2 + 72W$, so

$E(M) = (5)(2) + (10)(4) + (72)(1.5) = 158$ and

$\sigma_M^2 = (5)^2 (.5)^2 + (10)^2 (1)^2 + (72)^2 (.25)^2 = 430.25 \Rightarrow \sigma_M = 20.74$.

b. $P(M \le 200) = P\left(Z \le \dfrac{200 - 158}{20.74}\right) = P(Z \le 2.03) = .9788$.

73.

a. Both are approximately normal by the Central Limit Theorem.

b. The difference of two rvs is just an example of a linear combination, and a linear combination of normal rvs has a normal distribution, so $\overline{X} - \overline{Y}$ has approximately a normal distribution with $\mu_{\overline{X} - \overline{Y}} = 5$

and $\sigma_{\overline{X} - \overline{Y}} = \sqrt{\dfrac{8^2}{40} + \dfrac{6^2}{35}} = 1.621$.

c. $P(-1 \le \overline{X} - \overline{Y} \le 1) \approx P\left(\dfrac{-1 - 5}{1.6213} \le Z \le \dfrac{1 - 5}{1.6213}\right) = P(-3.70 \le Z \le -2.47) \approx .0068$.

d. $P(\bar{X} - \bar{Y} \geq 10) \approx P\left(Z \geq \dfrac{10-5}{1.6213}\right) = P(Z \geq 3.08) = .0010$. This probability is quite small, so such an occurrence is unlikely if $\mu_1 - \mu_2 = 5$, and we would thus doubt this claim.

Supplementary Exercises

75.

a. $p_X(x)$ is obtained by adding joint probabilities across the row labeled x, resulting in $p_X(x) = .2, .5, .3$ for $x = 12, 15, 20$ respectively. Similarly, from column sums $p_y(y) = .1, .35, .55$ for $y = 12, 15, 20$ respectively.

b. $P(X \leq 15 \text{ and } Y \leq 15) = p(12,12) + p(12,15) + p(15,12) + p(15,15) = .25$.

c. $p_X(12) \cdot p_Y(12) = (.2)(.1) \neq .05 = p(12,12)$, so X and Y are not independent. (Almost any other (x, y) pair yields the same conclusion).

d. $E(X + Y) = \sum\sum (x+y)p(x,y) = 33.35$ (or $= E(X) + E(Y) = 33.35$).

e. $E(|X - Y|) = \sum\sum |x - y| p(x,y) = \cdots = 3.85$.

77.

a. $1 = \int_{-\infty}^{\infty}\int_{-\infty}^{\infty} f(x,y)\,dxdy = \int_0^{20}\int_{20-x}^{30-x} kxy\,dydx + \int_{20}^{30}\int_0^{30-x} kxy\,dydx = \dfrac{81,250}{3} \cdot k \Rightarrow k = \dfrac{3}{81,250}$.

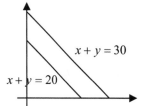

$x + y = 30$

$x + y = 20$

b. $f_X(x) = \begin{cases} \int_{20-x}^{30-x} kxy\,dy = k(250x - 10x^2) & 0 \leq x \leq 20 \\ \int_0^{30-x} kxy\,dy = k(450x - 30x^2 + \frac{1}{2}x^3) & 20 \leq x \leq 30 \end{cases}$

By symmetry, $f_Y(y)$ is obtained by substituting y for x in $f_X(x)$.
Since $f_X(25) > 0$ and $f_Y(25) > 0$, but $f(25, 25) = 0$, $f_X(x) \cdot f_Y(y) \neq f(x,y)$ for all (x, y), so X and Y are not independent.

c. $P(X + Y \leq 25) = \int_0^{20}\int_{20-x}^{25-x} kxy\,dydx + \int_{20}^{25}\int_0^{25-x} kxy\,dydx = \dfrac{3}{81,250} \cdot \dfrac{230,625}{24} = .355$.

d. $E(X + Y) = E(X) + E(Y) = 2E(X) = 2\left\{ \int_0^{20} x \cdot k\left(250x - 10x^2\right)dx \right.$

$\left. + \int_{20}^{30} x \cdot k\left(450x - 30x^2 + \frac{1}{2}x^3\right)dx \right\} = 2k(351,666.67) = 25.969$.

92

e. $E(XY) = \int_{-\infty}^{\infty} \int_{-\infty}^{\infty} xy \cdot f(x,y)dxdy = \int_{0}^{20} \int_{20-x}^{30-x} kx^2y^2 dydx$

$+ \int_{20}^{30} \int_{0}^{30-x} kx^2y^2 dydx = \frac{k}{3} \cdot \frac{33,250,000}{3} = 136.4103$, so

$\text{Cov}(X, Y) = 136.4103 - (12.9845)^2 = -32.19$.

$E(X^2) = E(Y^2) = 204.6154$, so $\sigma_X^2 = \sigma_Y^2 = 204.6154 - (12.9845)^2 = 36.0182$ and $\rho = \frac{-32.19}{36.0182} = -.894$.

f. $V(X + Y) = V(X) + V(Y) + 2\text{Cov}(X, Y) = 7.66$.

79. $E(\bar{X} + \bar{Y} + \bar{Z}) = 500 + 900 + 2000 = 3400$.

$V(\bar{X} + \bar{Y} + \bar{Z}) = \frac{50^2}{365} + \frac{100^2}{365} + \frac{180^2}{365} = 123.014 \Rightarrow SD(\bar{X} + \bar{Y} + \bar{Z}) = 11.09$.

$P(\bar{X} + \bar{Y} + \bar{Z} \le 3500) = P(Z \le 9.0) \approx 1$.

81.

a. $E(N) \cdot \mu = (10)(40) = 400$ minutes.

b. We expect 20 components to come in for repair during a 4 hour period,
so $E(N) \cdot \mu = (20)(3.5) = 70$.

83. $0.95 = P(\mu - .02 \le \bar{X} \le \mu + .02) = P\left(\frac{-.02}{.1/\sqrt{n}} \le Z \le \frac{.02}{.1/\sqrt{n}}\right) = P\left(-.2\sqrt{n} \le Z \le .2\sqrt{n}\right)$; since

$P(-1.96 \le Z \le 1.96) = .95$, $.2\sqrt{n} = 1.96 \Rightarrow n = 97$. The Central Limit Theorem justifies our use of the

normal distribution here.

85. The expected value and standard deviation of volume are 87,850 and 4370.37, respectively, so

$P(\text{volume} \le 100,000) = P\left(Z \le \frac{100,000 - 87,850}{4370.37}\right) = P(Z \le 2.78) = .9973$.

87.

a. $V(aX + Y) = a^2\sigma_X^2 + 2a\text{Cov}(X,Y) + \sigma_Y^2 = a^2\sigma_X^2 + 2a\sigma_X\sigma_Y\rho + \sigma_Y^2$.

Substituting $a = \frac{\sigma_Y}{\sigma_X}$ yields $\sigma_Y^2 + 2\sigma_Y^2\rho + \sigma_Y^2 = 2\sigma_Y^2(1 + \rho) \ge 0$. This implies $(1 + \rho) \ge 0$, or $\rho \ge -1$.

b. The same argument as in **a** yields $2\sigma_Y^2(1 - \rho) \ge 0$, from which $\rho \le 1$.

c. Suppose $\rho = 1$. Then $V(aX - Y) = 2\sigma_Y^2(1 - \rho) = 0$, which implies that $aX - Y$ is a constant. Solve for Y
and $Y = aX - (\text{constant})$, which is of the form $aX + b$.

89.

a. With $Y = X_1 + X_2$, $F_Y(y) = \int_0^y \left\{ \int_0^{y-x_1} \frac{1}{2^{v_1/2}\Gamma(v_1/2)} \cdot \frac{1}{2^{v_2/2}\Gamma(v_2/2)} \cdot x_1^{\frac{v_1}{2}-1} x_2^{\frac{v_2}{2}-1} e^{-\frac{x_1+x_2}{2}} dx_2 \right\} dx_1$. But the

inner integral can be shown to be equal to $\frac{1}{2^{(v_1+v_2)/2}\Gamma((v_1+v_2)/2)} y^{[(v_1+v_2)/2]-1} e^{-y/2}$, from which the result

follows.

b. By **a**, $Z_1^2 + Z_2^2$ is chi-squared with $v = 2$, so $(Z_1^2 + Z_2^2) + Z_3^2$ is chi-squared with $v = 3$, etc., until $Z_1^2 + \ldots + Z_n^2$ is chi-squared with $v = n$.

c. $\frac{X_i - \mu}{\sigma}$ is standard normal, so $\left[\frac{X_i - \mu}{\sigma}\right]^2$ is chi-squared with $v = 1$, so the sum is chi-squared with parameter $v = n$.

91.

a. $V(X_1) = V(W + E_1) = \sigma_W^2 + \sigma_E^2 = V(W + E_2) = V(X_2)$ and $\text{Cov}(X_1, X_2) =$
$\text{Cov}(W + E_1, W + E_2) = \text{Cov}(W,W) + \text{Cov}(W,E_2) + \text{Cov}(E_1,W) + \text{Cov}(E_1,E_2) =$
$\text{Cov}(W, W) + 0 + 0 + 0 = V(W) = \sigma_W^2$.

Thus, $\rho = \dfrac{\sigma_W^2}{\sqrt{\sigma_W^2 + \sigma_E^2} \cdot \sqrt{\sigma_W^2 + \sigma_E^2}} = \dfrac{\sigma_W^2}{\sigma_W^2 + \sigma_E^2}$.

b. $\rho = \dfrac{1}{1 + .0001} = .9999$.

93. $E(Y) \doteq h(\mu_1, \mu_2, \mu_3, \mu_4) = 120\left[\frac{1}{10} + \frac{1}{15} + \frac{1}{20}\right] = 26$.

The partial derivatives of $h(\mu_1, \mu_2, \mu_3, \mu_4)$ with respect to x_1, x_2, x_3, and x_4 are $-\dfrac{x_4}{x_1^2}$, $-\dfrac{x_4}{x_2^2}$, $-\dfrac{x_4}{x_3^2}$, and

$\dfrac{1}{x_1} + \dfrac{1}{x_2} + \dfrac{1}{x_3}$, respectively. Substituting $x_1 = 10$, $x_2 = 15$, $x_3 = 20$, and $x_4 = 120$ gives $-1.2, -.5333, -.3000$,
and $.2167$, respectively, so $V(Y) = (1)(-1.2)^2 + (1)(-.5333)^2 + (1.5)(-.3000)^2 + (4.0)(.2167)^2 = 2.6783$, and
the approximate sd of Y is 1.64.

95. Since X and Y are standard normal, each has mean 0 and variance 1.

a. $\text{Cov}(X, U) = \text{Cov}(X, .6X + .8Y) = .6\text{Cov}(X, X) + .8\text{Cov}(X, Y) = .6V(X) + .8(0) = .6(1) = .6$.
The covariance of X and Y is zero because X and Y are independent.
Also, $V(U) = V(.6X + .8Y) = (.6)^2V(X) + (.8)^2V(Y) = (.36)(1) + (.64)(1) = 1$. Therefore,

$$\text{Corr}(X,U) = \frac{\text{Cov}(X,U)}{\sigma_X \sigma_U} = \frac{.6}{\sqrt{1}\sqrt{1}} = .6, \text{ the coefficient on } X.$$

b. Based on part **a**, for any specified ρ we want $U = \rho X + bY$, where the coefficient b on Y has the feature
that $\rho^2 + b^2 = 1$ (so that the variance of U equals 1). One possible option for b is $b = \sqrt{1 - \rho^2}$, from
which $U = \rho X + \sqrt{1 - \rho^2} Y$.

94

CHAPTER 6

Section 6.1

1.

    a. We use the sample mean, $\bar{x}$, to estimate the population mean μ. $\hat{\mu} = \bar{x} = \dfrac{\Sigma x_i}{n} = \dfrac{219.80}{27} = 8.1407$.

    b. We use the sample median, $\tilde{x} = 7.7$ (the middle observation when arranged in ascending order).

    c. We use the sample standard deviation, $s = \sqrt{s^2} = \sqrt{\dfrac{1860.94 - \frac{(219.8)^2}{27}}{26}} = 1.660$.

    d. With "success" = observation greater than 10, x = # of successes = 4, and $\hat{p} = \dfrac{x}{n} = \dfrac{4}{27} = .1481$.

    e. We use the sample (std dev)/(mean), or $\dfrac{s}{\bar{x}} = \dfrac{1.660}{8.1407} = .2039$.

3.

    a. We use the sample mean, $\bar{x} = 1.3481$.

    b. Because we assume normality, the mean = median, so we also use the sample mean $\bar{x} = 1.3481$. We could also easily use the sample median.

    c. We use the 90th percentile of the sample: $\hat{\mu} + (1.28)\hat{\sigma} = \bar{x} + 1.28s = 1.3481 + (1.28)(.3385) = 1.7814$.

    d. Since we can assume normality,
$$P(X < 1.5) \approx P\left(Z < \frac{1.5 - \bar{x}}{s}\right) = P\left(Z < \frac{1.5 - 1.3481}{.3385}\right) = P(Z < .45) = .6736.$$

    e. The estimated standard error of $\bar{x} = \dfrac{\hat{\sigma}}{\sqrt{n}} = \dfrac{s}{\sqrt{n}} = \dfrac{.3385}{\sqrt{16}} = .0846$.

5.   Let θ = the total audited value. Three potential estimators of θ are $\hat{\theta}_1 = N\bar{X}$, $\hat{\theta}_2 = T - N\bar{D}$, and $\hat{\theta}_3 = T \cdot \dfrac{\bar{X}}{\bar{Y}}$.

From the data, $\bar{y} = 374.6$, $\bar{x} = 340.6$, and $\bar{d} = 34.0$. Knowing $N = 5{,}000$ and $T = 1{,}761{,}300$, the three corresponding estimates are $\hat{\theta}_1 = (5{,}000)(340.6) = 1{,}703{,}000$, $\hat{\theta}_2 = 1{,}761{,}300 - (5{,}000)(34.0) = 1{,}591{,}300$,

and $\hat{\theta}_3 = 1{,}761{,}300\left(\dfrac{340.6}{374.6}\right) = 1{,}601{,}438.281$.

7.

   a. $\hat{\mu} = \bar{x} = \dfrac{\sum x_i}{n} = \dfrac{1206}{10} = 120.6.$

   b. $\hat{\tau} = 10{,}000 \; \hat{\mu} = 1{,}206{,}000.$

   c. 8 of 10 houses in the sample used at least 100 therms (the "successes"), so $\hat{p} = \frac{8}{10} = .80.$

   d. The ordered sample values are 89, 99, 103, 109, 118, 122, 125, 138, 147, 156, from which the two middle values are 118 and 122, so $\hat{\tilde{\mu}} = \tilde{x} = \dfrac{118 + 122}{2} = 120.0.$

9.

   a. $E(\bar{X}) = \mu = E(X),$ so $\bar{X}$ is an unbiased estimator for the Poisson parameter μ. Since $n = 150$,

$$\hat{\mu} = \bar{x} = \frac{\Sigma x_i}{n} = \frac{(0)(18) + (1)(37) + \ldots + (7)(1)}{150} = \frac{317}{150} = 2.11.$$

   b. $\sigma_{\bar{x}} = \dfrac{\sigma}{\sqrt{n}} = \dfrac{\sqrt{\mu}}{\sqrt{n}}$, so the estimated standard error is $\sqrt{\dfrac{\hat{\mu}}{n}} = \dfrac{\sqrt{2.11}}{\sqrt{150}} = .119.$

11.

   a. $E\left(\dfrac{X_1}{n_1} - \dfrac{X_2}{n_2}\right) = \dfrac{1}{n_1}E(X_1) - \dfrac{1}{n_2}E(X_2) = \dfrac{1}{n_1}(n_1 p_1) - \dfrac{1}{n_2}(n_2 p_2) = p_1 - p_2.$

   b. $V\left(\dfrac{X_1}{n_1} - \dfrac{X_2}{n_2}\right) = V\left(\dfrac{X_1}{n_1}\right) + V\left(\dfrac{X_2}{n_2}\right) = \left(\dfrac{1}{n_1}\right)^2 V(X_1) + \left(\dfrac{1}{n_2}\right)^2 V(X_2) =$

$$\frac{1}{n_1^2}(n_1 p_1 q_1) + \frac{1}{n_2^2}(n_2 p_2 q_2) = \frac{p_1 q_1}{n_1} + \frac{p_2 q_2}{n_2},$$ and the standard error is the square root of this quantity.

   c. With $\hat{p}_1 = \dfrac{x_1}{n_1}$, $\hat{q}_1 = 1 - \hat{p}_1$, $\hat{p}_2 = \dfrac{x_2}{n_2}$, $\hat{q}_2 = 1 - \hat{p}_2$, the estimated standard error is $\sqrt{\dfrac{\hat{p}_1 \hat{q}_1}{n_1} + \dfrac{\hat{p}_2 \hat{q}_2}{n_2}}.$

   d. $(\hat{p}_1 - \hat{p}_2) = \dfrac{127}{200} - \dfrac{176}{200} = .635 - .880 = -.245$

   e. $\sqrt{\dfrac{(.635)(.365)}{200} + \dfrac{(.880)(.120)}{200}} = .041$

13. $\mu = E(X) = \displaystyle\int_{-1}^{1} x \cdot \frac{1}{2}(1 + \theta x)\,dx = \left. \frac{x^2}{4} + \frac{\theta x^3}{6} \right|_{-1}^{1} = \frac{1}{3}\theta \Rightarrow \theta = 3\mu \Rightarrow$

$\hat{\theta} = 3\bar{X} \Rightarrow E(\hat{\theta}) = E(3\bar{X}) = 3E(\bar{X}) = 3\mu = 3\left(\dfrac{1}{3}\right)\theta = \theta.$

15.

a. $E(X^2) = 2\theta$ implies that $E\left(\dfrac{X^2}{2}\right) = \theta$. Consider $\hat{\theta} = \dfrac{\sum X_i^2}{2n}$. Then

$$E(\hat{\theta}) = E\left(\dfrac{\sum X_i^2}{2n}\right) = \dfrac{\sum E(X_i^2)}{2n} = \dfrac{\sum 2\theta}{2n} = \dfrac{2n\theta}{2n} = \theta,$$ implying that $\hat{\theta}$ is an unbiased estimator for θ.

b. $\sum x_i^2 = 1490.1058$, so $\hat{\theta} = \dfrac{1490.1058}{20} = 74.505$.

17.

a. $E(\hat{p}) = \displaystyle\sum_{x=0}^{\infty} \dfrac{r-1}{x+r-1} \cdot \binom{x+r-1}{x} \cdot p^r \cdot (1-p)^x$

$= p \displaystyle\sum_{x=0}^{\infty} \dfrac{(x+r-2)!}{x!(r-2)!} \cdot p^{r-1} \cdot (1-p)^x = p \displaystyle\sum_{x=0}^{\infty} \binom{x+r-2}{x} p^{r-1} (1-p)^x = p \displaystyle\sum_{x=0}^{\infty} nb(x; r-1, p) = p$.

b. For the given sequence, $x = 5$, so $\hat{p} = \dfrac{5-1}{5+5-1} = \dfrac{4}{9} = .444$.

19.

a. $\lambda = .5p + .15 \Rightarrow 2\lambda = p + .3$, so $p = 2\lambda - .3$ and $\hat{p} = 2\hat{\lambda} - .3 = 2\left(\dfrac{Y}{n}\right) - .3$; the estimate is

$2\left(\dfrac{20}{80}\right) - .3 = .2$.

b. $E(\hat{p}) = E(2\hat{\lambda} - .3) = 2E(\hat{\lambda}) - .3 = 2\lambda - .3 = p$, as desired.

c. Here $\lambda = .7p + (.3)(.3)$, so $p = \dfrac{10}{7}\lambda - \dfrac{9}{70}$ and $\hat{p} = \dfrac{10}{7}\left(\dfrac{Y}{n}\right) - \dfrac{9}{70}$.

Section 6.2

21.

a. $E(X) = \beta \cdot \Gamma\left(1+\frac{1}{\alpha}\right)$ and $E(X^2) = V(X) + [E(X)]^2 = \beta^2\Gamma\left(1+\frac{2}{\alpha}\right)$, so the moment estimators $\hat{\alpha}$ and

$\hat{\beta}$ are the solution to $\bar{X} = \hat{\beta}\cdot\Gamma\left(1+\frac{1}{\hat{\alpha}}\right)$, $\frac{1}{n}\sum X_i^2 = \hat{\beta}^2\Gamma\left(1+\frac{2}{\hat{\alpha}}\right)$. Thus $\hat{\beta} = \dfrac{\bar{X}}{\Gamma\left(1+\dfrac{1}{\hat{\alpha}}\right)}$, so once $\hat{\alpha}$

has been determined $\Gamma\left(1+\frac{1}{\hat{\alpha}}\right)$ is evaluated and $\hat{\beta}$ then computed. Since $\bar{X}^2 = \hat{\beta}^2\cdot\Gamma^2\left(1+\frac{1}{\hat{\alpha}}\right)$,

$\dfrac{1}{n}\sum\dfrac{X_i^2}{\bar{X}^2} = \dfrac{\Gamma\left(1+\dfrac{2}{\hat{\alpha}}\right)}{\Gamma^2\left(1+\dfrac{1}{\hat{\alpha}}\right)}$, so this equation must be solved to obtain $\hat{\alpha}$.

b. From a, $\dfrac{1}{20}\left(\dfrac{16,500}{28.0^2}\right) = 1.05 = \dfrac{\Gamma\left(1+\dfrac{2}{\hat{\alpha}}\right)}{\Gamma^2\left(1+\dfrac{1}{\hat{\alpha}}\right)}$, so $\dfrac{1}{1.05} = .95 = \dfrac{\Gamma^2\left(1+\dfrac{1}{\hat{\alpha}}\right)}{\Gamma\left(1+\dfrac{2}{\hat{\alpha}}\right)}$, and from the hint,

$\dfrac{1}{\hat{\alpha}} = .2 \Rightarrow \hat{\alpha} = 5$. Then $\hat{\beta} = \dfrac{\bar{x}}{\Gamma(1.2)} = \dfrac{28.0}{\Gamma(1.2)}$.

23. For a single sample from a Poisson distribution, $f(x_1,...,x_n;\mu) = \dfrac{e^{-\mu}\mu^{x_1}}{x_1!}\cdots\dfrac{e^{-\mu}\mu^{x_n}}{x_n!} = \dfrac{e^{-n\mu}\mu^{\Sigma x_i}}{x_1!\cdots x_n!}$, so

$\ln\left[f(x_1,...,x_n;\mu)\right] = -n\mu + \sum x_i \ln(\mu) - \sum \ln(x_i!)$. Thus

$\dfrac{d}{d\mu}\left[\ln\left[f(x_1,...,x_n;\lambda)\right]\right] = -n + \dfrac{\sum x_i}{\mu} = 0 \Rightarrow \hat{\mu} = \dfrac{\sum x_i}{n} = \bar{x}$. For our problem, $f(x_1,...,x_n,y_1...y_n;\mu_1,\mu_2)$ is

a product of the X sample likelihood and the Y sample likelihood, implying that $\hat{\mu}_1 = \bar{x}, \hat{\mu}_2 = \bar{y}$, and (by the

invariance principle) $\widehat{(\mu_1 - \mu_2)} = \bar{x} - \bar{y}$.

25.

a. $\hat{\mu} = \bar{x} = 384.4; s^2 = 395.16$, so $\dfrac{1}{n}\sum(x_i - \bar{x})^2 = \hat{\sigma}^2 = \dfrac{9}{10}(395.16) = 355.64$ and $\hat{\sigma} = \sqrt{355.64} = 18.86$

(this is not s).

b. The 95th percentile is $\mu + 1.645\sigma$, so the mle of this is (by the invariance principle)

$\hat{\mu} + 1.645\hat{\sigma} = 415.42$.

27.

a. $f(x_1,\ldots,x_n;\alpha,\beta) = \dfrac{(x_1 x_2 \ldots x_n)^{\alpha-1} e^{-\Sigma x_i/\beta}}{\beta^{n\alpha} \Gamma^n(\alpha)}$, so the log likelihood is

$(\alpha - 1)\sum \ln(x_i) - \dfrac{\sum x_i}{\beta} - n\alpha \ln(\beta) - n \ln \Gamma(\alpha)$. Equating both $\dfrac{d}{d\alpha}$ and $\dfrac{d}{d\beta}$ to 0 yields

$\sum \ln(x_i) - n\ln(\beta) - n\dfrac{d}{d\alpha}\Gamma(\alpha) = 0$ and $\dfrac{\sum x_i}{\beta^2} - \dfrac{n\alpha}{\beta} = 0$, a very difficult system of equations to solve.

b. From the second equation in **a**, $\dfrac{\sum x_i}{\beta} = n\alpha \Rightarrow \bar{x} = \alpha\beta = \mu$, so the mle of μ is $\hat{\mu} = \bar{X}$.

29.

a. The joint pdf (likelihood function) is

$$f(x_1,\ldots,x_n;\lambda,\theta) = \begin{cases} \lambda^n e^{-\lambda\Sigma(x_i-\theta)} & x_1 \geq \theta,\ldots,x_n \geq \theta \\ 0 & \text{otherwise} \end{cases}$$

Notice that $x_1 \geq \theta,\ldots,x_n \geq \theta$ iff $\min(x_i) \geq \theta$, and that $-\lambda\Sigma(x_i - \theta) = -\lambda\Sigma x_i + n\lambda\theta$.

Thus likelihood $= \begin{cases} \lambda^n \exp(-\lambda\Sigma x_i)\exp(n\lambda\theta) & \min(x_i) \geq \theta \\ 0 & \min(x_i) < \theta \end{cases}$

Consider maximization with respect to θ. Because the exponent $n\lambda\theta$ is positive, increasing θ will increase the likelihood provided that $\min(x_i) \geq \theta$; if we make θ larger than $\min(x_i)$, the likelihood drops to 0. This implies that the mle of θ is $\hat{\theta} = \min(x_i)$. The log likelihood is now

$n\ln(\lambda) - \lambda\Sigma(x_i - \hat{\theta})$. Equating the derivative w.r.t. λ to 0 and solving yields $\hat{\lambda} = \dfrac{n}{\Sigma(x_i - \hat{\theta})} = \dfrac{n}{\Sigma x_i - n\hat{\theta}}$.

b. $\hat{\theta} = \min(x_i) = .64$, and $\Sigma x_i = 55.80$, so $\hat{\lambda} = \dfrac{10}{55.80 - 6.4} = .202$.

Supplementary Exercises

31. Substitute $k = \varepsilon/\sigma_Y$ into Chebyshev's inequality to write $P(|Y - \mu_Y| \geq \varepsilon) \leq 1/(\varepsilon/\sigma_Y)^2 = V(Y)/\varepsilon^2$. Since $E(\bar{X}) = \mu$ and $V(\bar{X}) = \sigma^2/n$, we may then write $P(|\bar{X} - \mu| \geq \varepsilon) \leq \dfrac{\sigma^2/n}{\varepsilon^2}$. As $n \to \infty$, this fraction converges to 0, hence $P(|\bar{X} - \mu| \geq \varepsilon) \to 0$, as desired.

33. Let x_1 = the time until the first birth, x_2 = the elapsed time between the first and second births, and so on. Then $f(x_1,\ldots,x_n;\lambda) = \lambda e^{-\lambda x_1} \cdot (2\lambda)e^{-2\lambda x_2}\ldots(n\lambda)e^{-n\lambda x_n} = n!\lambda^n e^{-\lambda\Sigma k x_k}$. Thus the log likelihood is

$\ln(n!) + n\ln(\lambda) - \lambda\Sigma k x_k$. Taking $\dfrac{d}{d\lambda}$ and equating to 0 yields $\hat{\lambda} = \dfrac{n}{\Sigma k x_k}$.

For the given sample, $n = 6$, $x_1 = 25.2$, $x_2 = 41.7 - 25.2 = 16.5$, $x_3 = 9.5$, $x_4 = 4.3$, $x_5 = 4.0$, $x_6 = 2.3$; so

$\displaystyle\sum_{k=1}^{6} kx_k = (1)(25.2) + (2)(16.5) + \ldots + (6)(2.3) = 137.7$ and $\hat{\lambda} = \dfrac{6}{137.7} = .0436$.

35.

$x_i + x_j$	23.5	26.3	28.0	28.2	29.4	29.5	30.6	31.6	33.9	49.3
23.5	23.5	24.9	25.75	25.85	26.45	26.5	27.05	27.55	28.7	36.4
26.3		26.3	27.15	27.25	27.85	27.9	28.45	28.95	30.1	37.8
28.0			28.0	28.1	28.7	28.75	29.3	29.8	30.95	38.65
28.2				28.2	28.8	28.85	29.4	29.9	31.05	38.75
29.4					29.4	29.45	30.0	30.5	30.65	39.35
29.5						29.5	30.05	30.55	31.7	39.4
30.6							30.6	31.1	32.25	39.95
31.6								31.6	32.75	40.45
33.9									33.9	41.6
49.3										49.3

There are 55 averages, so the median is the 28th in order of increasing magnitude. Therefore, $\hat{\mu} = 29.5$.

37. Let $c = \dfrac{\Gamma\left(\frac{n-1}{2}\right)}{\Gamma\left(\frac{n}{2}\right) \cdot \sqrt{\frac{2}{n-1}}}$. Then $E(cS) = cE(S)$, and c cancels with the two Γ factors and the square root in $E(S)$,

leaving just σ. When $n = 20$, $c = \dfrac{\Gamma(9.5)}{\Gamma(10) \cdot \sqrt{\frac{2}{19}}} = \dfrac{(8.5)(7.5)\cdots(.5)\Gamma(.5)}{(10-1)!\sqrt{\frac{2}{19}}} = \dfrac{(8.5)(7.5)\cdots(.5)\sqrt{\pi}}{9!\sqrt{\frac{2}{19}}} = 1.0132$.

CHAPTER 7

Section 7.1

1.

    a. $z_{\alpha/2} = 2.81$ implies that $\alpha/2 = 1 - \Phi(2.81) = .0025$, so $\alpha = .005$ and the confidence level is $100(1-\alpha)\% = 99.5\%$.

    b. $z_{\alpha/2} = 1.44$ implies that $\alpha = 2[1 - \Phi(1.44)] = .15$, and the confidence level is $100(1-\alpha)\% = 85\%$.

    c. 99.7% confidence implies that $\alpha = .003$, $\alpha/2 = .0015$, and $z_{.0015} = 2.96$. (Look for cumulative area equal to $1 - .0015 = .9985$ in the main body of table A.3.) Or, just use $z \approx 3$ by the empirical rule.

    d. 75% confidence implies $\alpha = .25$, $\alpha/2 = .125$, and $z_{.125} = 1.15$.

3.

    a. A 90% confidence interval will be narrower. The z critical value for a 90% confidence level is 1.645, smaller than the z of 1.96 for the 95% confidence level, thus producing a narrower interval.

    b. Not a correct statement. Once and interval has been created from a sample, the mean μ is either enclosed by it, or not. We have 95% confidence in the general procedure, under repeated and independent sampling.

    c. Not a correct statement. The interval is an estimate for the population mean, not a boundary for population values.

    d. Not a correct statement. In theory, if the process were repeated an infinite number of times, 95% of the intervals would contain the population mean μ. We *expect* 95 out of 100 intervals will contain μ, but we don't know this to be true.

5.

    a. $4.85 \pm \dfrac{(1.96)(.75)}{\sqrt{20}} = 4.85 \pm .33 = (4.52, 5.18)$.

    b. $z_{\alpha/2} = z.01 = 2.33$, so the interval is $4.56 \pm \dfrac{(2.33)(.75)}{\sqrt{16}} = (4.12, 5.00)$.

    c. $n = \left[\dfrac{2(1.96)(.75)}{.40} \right]^2 = 54.02 \nearrow 55$.

    d. Width $w = 2(.2) = .4$, so $n = \left[\dfrac{2(2.58)(.75)}{.4} \right]^2 = 93.61 \nearrow 94$.

7. If $L = 2z_{\alpha/2} \dfrac{\sigma}{\sqrt{n}}$ and we increase the sample size by a factor of 4, the new length is

$L' = 2z_{\alpha/2} \dfrac{\sigma}{\sqrt{4n}} = \left[2z_{\alpha/2} \dfrac{\sigma}{\sqrt{n}} \right]\left(\dfrac{1}{2} \right) = \dfrac{L}{2}$. Thus halving the length requires n to be increased fourfold. If

$n' = 25n$, then $L' = \dfrac{L}{5}$, so the length is decreased by a factor of 5.

9.

    a. $\left(\bar{x} - 1.645 \dfrac{\sigma}{\sqrt{n}}, \infty \right)$. From 5a, $\bar{x} = 4.85$, $\sigma = .75$, and $n = 20$; $4.85 - 1.645 \dfrac{.75}{\sqrt{20}} = 4.5741$, so the

    interval is $(4.5741, \infty)$.

    b. $\left(\bar{x} - z_{\alpha} \dfrac{\sigma}{\sqrt{n}}, \infty \right)$

    c. $\left(-\infty, \bar{x} + z_{\alpha} \dfrac{\sigma}{\sqrt{n}} \right)$; From 4a, $\bar{x} = 58.3$, $\sigma = 3.0$, and $n = 25$; $58.3 + 2.33 \dfrac{3}{\sqrt{25}} = 59.70$, so the interval is

    $(-\infty, 59.70)$.

11. Y is a binomial rv with $n = 1000$ and $p = .95$, so $E(Y) = np = 950$, the expected number of intervals that capture μ, and $\sigma_Y = \sqrt{npq} = 6.892$. Using the normal approximation to the binomial distribution, $P(940 \le Y \le 960) = P(939.5 \le Y \le 960.5) \approx P(-1.52 \le Z \le 1.52) = .9357 - .0643 = .8714$.

Section 7.2

13.

    a. $\bar{x} \pm z_{.025} \dfrac{s}{\sqrt{n}} = 654.16 \pm 1.96 \dfrac{164.43}{\sqrt{50}} = (608.58, 699.74)$. We are 95% confident that the true average

    CO_2 level in this population of homes with gas cooking appliances is between 608.58ppm and
    699.74ppm

    b. $w = 50 = \dfrac{2(1.96)(175)}{\sqrt{n}} \Rightarrow \sqrt{n} = \dfrac{2(1.96)(175)}{50} = 13.72 \Rightarrow n = (13.72)^2 = 188.24$, which rounds up to 189.

15.

    a. $z_{\alpha} = .84$, and $\Phi(.84) = .7995 \approx .80$, so the confidence level is 80%.

    b. $z_{\alpha} = 2.05$, and $\Phi(2.05) = .9798 \approx .98$, so the confidence level is 98%.

    c. $z_{\alpha} = .67$, and $\Phi(.67) = .7486 \approx .75$, so the confidence level is 75%.

17. $\bar{x} - z_{.01}\dfrac{s}{\sqrt{n}} = 135.39 - 2.33\dfrac{4.59}{\sqrt{153}} = 135.39 - .865 = 134.53$. We are 99% confident that the true average ultimate tensile strength is greater than 134.53.

19. $\hat{p} = \dfrac{201}{356} = .5646$; We calculate a 95% confidence interval for the proportion of all dies that pass the probe:

$$\dfrac{.5646 + \dfrac{(1.96)^2}{2(356)} \pm 1.96\sqrt{\dfrac{(.5646)(.4354)}{356} + \dfrac{(1.96)^2}{4(356)^2}}}{1 + \dfrac{(1.96)^2}{356}} = \dfrac{.5700 \pm .0518}{1.01079} = (.513, .615).$$ The simpler CI formula

(7.11) gives $.5646 \pm 1.96\sqrt{\dfrac{.5646(.4354)}{356}} = (.513, .616)$, which is almost identical.

21. For a one-sided bound, we need $z_\alpha = z_{.05} = 1.645$; $\hat{p} = \dfrac{250}{1000} = .25$; and $\tilde{p} = \dfrac{.25 + 1.645^2/2000}{1 + 1.645^2/1000} = .2507$. The resulting 95% upper confidence bound for p, the true proportion of such consumers who never apply for a rebate, is $.2507 + \dfrac{1.645\sqrt{(.25)(.75)/1000 + (1.645)^2/(4 \cdot 1000^2)}}{1 + (1.645)^2/1000} = .2507 + .0225 = .2732$.

Yes, there is compelling evidence the true proportion is less than 1/3 (.3333), since we are 95% confident this true proportion is less than .2732.

23.

a. With such a large sample size, we can use the "simplified" CI formula (7.11). With $\hat{p} = .25$, $n = 2003$, and $z_{\alpha/2} = z_{.005} = 2.576$, the 99% confidence interval for p is

$$\hat{p} \pm z_{\alpha/2}\sqrt{\dfrac{\hat{p}\hat{q}}{n}} = .25 \pm 2.576\sqrt{\dfrac{(.25)(.75)}{2003}} = .25 \pm .025 = (.225, .275).$$

b. Using the "simplified" formula for sample size and $\hat{p} = \hat{q} = .5$,

$$n = \dfrac{4z^2\hat{p}\hat{q}}{w^2} = \dfrac{4(2.576)^2(.5)(.5)}{(.05)^2} = 2654.31$$

So, a sample of size at least 2655 is required. (We use $\hat{p} = \hat{q} = .5$ here, rather than the values from the sample data, so that our CI has the desired width irrespective of what the true value of p might be. See the textbook discussion toward the end of Section 7.2.)

25.

a. $n = \dfrac{2(1.96)^2(.25) - (1.96)^2(.01) \pm \sqrt{4(1.96)^4(.25)(.25 - .01) + .01(1.96)^4}}{.01} \approx 381$

b. $n = \dfrac{2(1.96)^2\left(\frac{1}{3} \cdot \frac{2}{3}\right) - (1.96)^2(.01) \pm \sqrt{4(1.96)^4\left(\frac{1}{3} \cdot \frac{2}{3}\right)\left(\frac{1}{3} \cdot \frac{2}{3} - .01\right) + .01(1.96)^4}}{.01} \approx 339$

27.   Note that the midpoint of the new interval is $\dfrac{x+z^2/2}{n+z^2}$, which is roughly $\dfrac{x+2}{n+4}$ with a confidence level of 95% and approximating $1.96 \approx 2$. The variance of this quantity is $\dfrac{np(1-p)}{\left(n+z^2\right)^2}$, or roughly $\dfrac{p(1-p)}{n+4}$. Now

replacing p with $\dfrac{x+2}{n+4}$, we have $\left(\dfrac{x+2}{n+4}\right) \pm z_{\alpha/2} \sqrt{\dfrac{\left(\dfrac{x+2}{n+4}\right)\left(1-\dfrac{x+2}{n+4}\right)}{n+4}}$. For clarity, let $x^* = x+2$ and

$n^* = n+4$, then $\hat{p}^* = \dfrac{x^*}{n^*}$ and the formula reduces to $\hat{p}^* \pm z_{\alpha/2} \sqrt{\dfrac{\hat{p}^*\hat{q}^*}{n^*}}$, the desired conclusion. For further discussion, see the Agresti article.

Section 7.3

29.

a. $t_{.025,10} = 2.228$ d. $t_{.005,50} = 2.678$

b. $t_{.025,20} = 2.086$ e. $t_{.01,25} = 2.485$

c. $t_{.005,20} = 2.845$ f. $-t_{.025,5} = -2.571$

31.

a. $t_{.05,10} = 1.812$ d. $t_{.01,4} = 3.747$

b. $t_{.05,15} = 1.753$ e. $t_{.02,24} \approx t_{.025,24} = 2.064$

c. $t_{.01,15} = 2.602$ f. $t_{.01,37} \approx 2.429$

33.

a. The boxplot indicates a very slight positive skew, with no outliers. The data appears to center near 438.

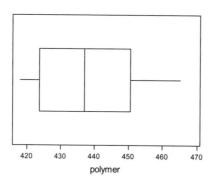

b. Based on a normal probability plot, it is reasonable to assume the sample observations came from a normal distribution.

c. With df $= n - 1 = 16$, the critical value for a 95% CI is $t_{.025,16} = 2.120$, and the interval is

$$438.29 \pm (2.120)\left(\frac{15.14}{\sqrt{17}}\right) = 438.29 \pm 7.785 = (430.51, 446.08).$$ Since 440 is within the interval, 440 is

a plausible value for the true mean. 450, however, is not, since it lies outside the interval.

35. $n = 15$, $\bar{x} = 25.0$, $s = 3.5$; $t_{.025,14} = 2.145$

  a. A 95% CI for the mean: $25.0 \pm 2.145\dfrac{3.5}{\sqrt{15}} = (23.06, 26.94)$.

  b. A 95% prediction interval: $25.0 \pm 2.145(3.5)\sqrt{1 + \dfrac{1}{15}} = (17.25, 32.75)$. The prediction interval is about 4 times wider than the confidence interval.

37.

  a. A 95% CI : $.9255 \pm 2.093(.0181) = .9255 \pm .0379 \Rightarrow (.8876, .9634)$

  b. A 95% P.I. : $.9255 \pm 2.093(.0809)\sqrt{1 + \frac{1}{20}} = .9255 \pm .1735 \Rightarrow (.7520, 1.0990)$

  c. A tolerance interval is requested, with $k = 99$, confidence level 95%, and $n = 20$. The tolerance critical value, from Table A.6, is 3.615. The interval is $.9255 \pm 3.615(.0809) \Rightarrow (.6330, 1.2180)$.

39.

  a. Based on the plot, generated by Minitab, it is plausible that the population distribution is normal.

Normal Probability Plot

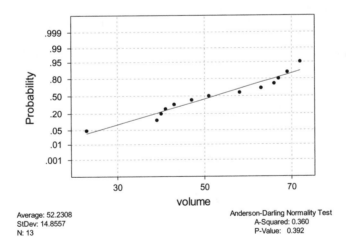

Average: 52.2308
StDev: 14.8557
N: 13

Anderson-Darling Normality Test
A-Squared: 0.360
P-Value: 0.392

  b. We require a tolerance interval. From table A.6, with 95% confidence, $k = 95$, and $n = 13$, the tolerance critical value is 3.081. $\bar{x} \pm 3.081s = 52.231 \pm 3.081(14.856) = 52.231 \pm 45.771 \Rightarrow (6.460, 98.002)$.

  c. A prediction interval, with $t_{.025,12} = 2.179$:
  $$52.231 \pm 2.179(14.856)\sqrt{1 + \frac{1}{13}} = 52.231 \pm 33.593 \Rightarrow (18.638, 85.824)$$

41. The 20 df row of Table A.5 shows that 1.725 captures upper tail area .05 and 1.325 captures upper tail area .10 The confidence level for each interval is 100(central area)%.
For the first interval, central area = 1 – sum of tail areas = 1 – (.25 + .05) = .70, and for the second and third intervals the central areas are 1 – (.20 + .10) = .70 and 1 – (.15 + .15) = 70. Thus each interval has

confidence level 70%. The width of the first interval is $\dfrac{s(.687+1.725)}{\sqrt{n}} = 2.412\dfrac{s}{\sqrt{n}}$, whereas the widths of

the second and third intervals are 2.185 and 2.128 standard errors respectively. The third interval, with symmetrically placed critical values, is the shortest, so it should be used. This will always be true for a t interval.

Section 7.4

43.

     a. $\chi^2_{.05,10} = 18.307$

     b. $\chi^2_{.95,10} = 3.940$

     c. Since $10.987 = \chi^2_{.975,22}$ and $36.78 = \chi^2_{.025,22}$, $P\left(\chi^2_{.975,22} \le \chi^2 \le \chi^2_{.025,22}\right) = .95$.

     d. Since $14.611 = \chi^2_{.95,25}$ and $37.652 = \chi^2_{.05,25}$, $P(\chi^2 < 14.611 \text{ or } \chi^2 > 37.652) =$
     $1 - P(\chi^2 > 14.611) + P(\chi^2 > 37.652) = (1 - .95) + .05 = .10$.

45. $n = 22$ implies that $df = n - 1 = 21$, so the .995 and .005 columns of Table A.7 give the necessary chi-squared critical values as 8.033 and 41.399. $\Sigma x_i = 1701.3$ and $\Sigma x_i^2 = 132,097.35$, so $s^2 = 25.368$. The

interval for σ^2 is $\left(\dfrac{21(25.368)}{41.399}, \dfrac{21(25.368)}{8.033}\right) = (12.868, 66.317)$ and the CI for σ is (3.6, 8.1). Validity of

this interval requires that fracture toughness be (at least approximately) normally distributed.

Supplementary Exercises

47.

     a. $n = 48$, $\bar{x} = 8.079a$, $s^2 = 23.7017$, and $s = 4.868$.
     A 95% CI for μ = the true average strength is

     $\bar{x} \pm 1.96\dfrac{s}{\sqrt{n}} = 8.079 \pm 1.96\dfrac{4.868}{\sqrt{48}} = 8.079 \pm 1.377 = (6.702, 9.456)$.

     b. $\hat{p} = \dfrac{13}{48} = .2708$. A 95% CI for p is

     $\dfrac{.2708 + \dfrac{1.96^2}{2(48)} \pm 1.96\sqrt{\dfrac{(.2708)(.7292)}{48} + \dfrac{1.96^2}{4(48)^2}}}{1 + \dfrac{1.96^2}{48}} = \dfrac{.3108 \pm .1319}{1.0800} = (.166, .410)$

49.

  a. There appears to be a slight positive skew in the middle half of the sample, but the lower whisker is much longer than the upper whisker. The extent of variability is rather substantial, although there are no outliers.

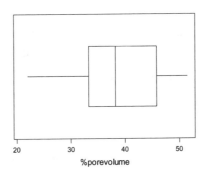

%porevolume

  b. The pattern of points in a normal probability plot is reasonably linear, so, yes, normality is plausible.

  c. $n = 18$, $\bar{x} = 38.66$, $s = 8.473$, and $t_{.01,17} = 2.567$. The 98% confidence interval is

$$38.66 \pm 2.567 \frac{8.473}{\sqrt{18}} = 38.66 \pm 5.13 = (33.53, 43.79).$$

51.

  a. With such a large sample size, we'll use the "simplified" CI formula (7.11). Here, $\hat{p} = \dfrac{1262}{2253} = .56$, so

a 95% CI for p is $\hat{p} \pm 1.96 \sqrt{\dfrac{\hat{p}\hat{q}}{n}} = .56 \pm 1.96 \sqrt{\dfrac{(.56)(.44)}{2253}} = .56 \pm .021 = (.539, .581)$. We are 95% confident that between 53.9% and 58.1% of all American adults have at some point used wireless means for online access.

  b. Using the "simplified" formula again, $n = \dfrac{4z^2 \hat{p}\hat{q}}{w^2} = \dfrac{4(1.96)^2(.5)(.5)}{(.04)^2} = 2401$. So, roughly 2400 people should be surveyed to assure a width no more than .04 with 95% confidence. (Note: using equation (7.12) gives $n = 2398$.)

  c. No. The upper bound in (a) uses a z-value of $1.96 = z_{.025}$. So, if this is used as an upper bound (and hence .025 equals α rather than $\alpha/2$), it gives a $(1 - .025) = 97.5\%$ upper bound. If we want a 95% confidence upper bound for p, 1.96 should be replaced by the critical value $z_{.05} = 1.645$.

53. With $\hat{\theta} = \frac{1}{3}(\bar{X}_1 + \bar{X}_2 + \bar{X}_3) - \bar{X}_4$, $\sigma_{\hat{\theta}}^2 = \frac{1}{9} V(\bar{X}_1 + \bar{X}_2 + \bar{X}_3) + V(\bar{X}_4) = \frac{1}{9}\left(\dfrac{\sigma_1^2}{n_1} + \dfrac{\sigma_2^2}{n_2} + \dfrac{\sigma_3^2}{n_3} \right) + \dfrac{\sigma_4^2}{n_4}$; $\hat{\sigma}_{\hat{\theta}}$ is

obtained by replacing each σ_i^2 by s_i^2 and taking the square root. The large-sample interval for θ is then

$$\frac{1}{3}(\bar{x}_1 + \bar{x}_2 + \bar{x}_3) - \bar{x}_4 \pm z_{\alpha/2} \sqrt{ \frac{1}{9}\left(\frac{s_1^2}{n_1} + \frac{s_2^2}{n_2} + \frac{s_3^2}{n_3} \right) + \frac{s_4^2}{n_4} }.$$

For the given data, $\hat{\theta} = -.50$ and $\hat{\sigma}_{\hat{\theta}} = .1718$, so the interval is $-.50 \pm 1.96(.1718) = (-.84, -.16)$.

55. The specified condition is that the interval be length .2, so $n = \left[\dfrac{2(1.96)(.8)}{.2} \right]^2 = 245.86 \nearrow 246$.

57. Proceeding as in Example 7.5 with T_r replacing ΣX_i, the CI for $\dfrac{1}{\lambda}$ is $\left(\dfrac{2t_r}{\chi^2_{1-\alpha/2,2r}}, \dfrac{2t_r}{\chi^2_{\alpha/2,2r}} \right)$ where

$t_r = y_1 + \ldots + y_r + (n-r) y_r$. In Example 6.7, $n = 20$, $r = 10$, and $t_r = 1115$. With df $= 20$, the necessary critical values are 9.591 and 34.170, giving the interval (65.3, 232.5). This is obviously an extremely wide interval. The censored experiment provides less information about $\dfrac{1}{\lambda}$ than would an uncensored experiment with $n = 20$.

59.

   a. $\displaystyle\int_{(\alpha/2)^{1/n}}^{(1-\alpha/2)^{1/n}} n u^{n-1} du = u^n \Big]_{(\alpha/2)^{1/n}}^{(1-\alpha/2)^{1/n}} = 1 - \dfrac{\alpha}{2} - \dfrac{\alpha}{2} = 1 - \alpha$. From the probability statement,

$\dfrac{(\alpha/2)^{1/n}}{\max(X_i)} \le \dfrac{1}{\theta} \le \dfrac{(1-\alpha/2)^{1/n}}{\max(X_i)}$ with probability $1 - \alpha$, so taking the reciprocal of each endpoint and

interchanging gives the CI $\left(\dfrac{\max(X_i)}{(1-\alpha/2)^{1/n}}, \dfrac{\max(X_i)}{(\alpha/2)^{1/n}} \right)$ for θ.

   b. $\alpha^{1/n} \le \dfrac{\max(X_i)}{\theta} \le 1$ with probability $1 - \alpha$, so $1 \le \dfrac{\theta}{\max(X_i)} \le \dfrac{1}{\alpha^{1/n}}$ with probability $1 - \alpha$, which yields

the interval $\left(\max(X_i), \dfrac{\max(X_i)}{\alpha^{1/n}} \right)$.

   c. It is easily verified that the interval of **b** is shorter — draw a graph of $f_U(u)$ and verify that the shortest interval which captures area $1 - \alpha$ under the curve is the rightmost such interval, which leads to the CI of **b**. With $\alpha = .05$, $n = 5$, and $\max(x_i) = 4.2$, this yields (4.2, 7.65).

61. $\tilde{x} = 76.2$, the lower and upper fourths are 73.5 and 79.7, respectively, and $f_s = 6.2$. The robust interval is

$76.2 \pm (1.93) \left(\dfrac{6.2}{\sqrt{22}} \right) = 76.2 \pm 2.6 = (73.6, 78.8)$.

$\bar{x} = 77.33$, $s = 5.037$, and $t_{.025,21} = 2.080$, so the t interval is

$77.33 \pm (2.080) \left(\dfrac{5.037}{\sqrt{22}} \right) = 77.33 \pm 2.23 = (75.1, 79.6)$. The t interval is centered at $\bar{x}$, which is pulled out

to the right of $\tilde{x}$ by the single mild outlier 93.7; the interval widths are comparable.

CHAPTER 8

Section 8.1

1.
    a. Yes. It is an assertion about the value of a parameter.

    b. No. The sample median $\tilde{X}$ is not a parameter.

    c. No. The sample standard deviation s is not a parameter.

    d. Yes. The assertion is that the standard deviation of population #2 exceeds that of population #1.

    e. No. $\bar{X}$ and $\bar{Y}$ are statistics rather than parameters, so they cannot appear in a hypothesis.

    f. Yes. H is an assertion about the value of a parameter.

3.   In this formulation, H_0 states the welds do not conform to specification. This assertion will not be rejected unless there is strong evidence to the contrary. Thus the burden of proof is on those who wish to assert that the specification is satisfied. Using H_a: $\mu < 100$ results in the welds being believed in conformance unless proved otherwise, so the burden of proof is on the non-conformance claim.

5.   Let σ denote the population standard deviation. The appropriate hypotheses are H_0: $\sigma = .05$ v. H_a: $\sigma < .05$. With this formulation, the burden of proof is on the data to show that the requirement has been met (the sheaths will not be used unless H_0 can be rejected in favor of H_a. Type I error: Conclude that the standard deviation is $< .05$ mm when it is really equal to .05 mm. Type II error: Conclude that the standard deviation is .05 mm when it is really $< .05$.

7.   A type I error here involves saying that the plant is not in compliance when in fact it is. A type II error occurs when we conclude that the plant is in compliance when in fact it isn't. Reasonable people may disagree as to which of the two errors is more serious. If in your judgment it is the type II error, then the reformulation H_0: $\mu = 150$ v. H_a: $\mu < 150$ makes the type I error more serious.

9.
    a. R_1 is most appropriate, because x either too large or too small contradicts $p = .5$ and supports $p \neq .5$.

    b. A type I error consists of judging one of the two companies favored over the other when in fact there is a 50-50 split in the population. A type II error involves judging the split to be 50-50 when it is not.

    c. X has a binomial distribution with $n = 25$ and $p = 0.5$.
        $\alpha = P(\text{type I error}) = P(X \leq 7 \text{ or } X \geq 18 \text{ when } X \sim \text{Bin}(25, .5)) = B(7; 25,.5) + [1 - B(17; 25,.5)] = .044$.

    d. $\beta(.4) = P(8 \leq X \leq 17 \text{ when } p = .4) = B(17; 25,.4) - B(7, 25,.4) = .845$; $\beta(.6) = .845$ also. Similarly, $\beta(.3) = B(17; 25, .3) - B(7; 25, .3) = .488 = \beta(.7)$.

    e. $x = 6$ is in the rejection region R_1, so H_0 is rejected in favor of H_a.

109

11.

    a. $H_0: \mu = 10$ v. $H_a: \mu \neq 10$.

    b. $\alpha = P(\text{rejecting } H_0 \text{ when } H_0 \text{ is true}) = P(\overline{X} \geq 10.1032 \text{ or } \overline{X} \leq 9.8968 \text{ when } \mu = 10)$.

        Since $\overline{X}$ is normally distributed with standard deviation $\frac{\sigma}{\sqrt{n}} = \frac{.2}{5} = .04$, $\alpha = P(Z \geq 2.58 \text{ or } Z \leq -2.58) =$

        $.005 + .005 = .01$.

    c. When $\mu = 10.1$, $E(\overline{X}) = 10.1$, so $\beta(10.1) = P(9.8968 < \overline{X} < 10.1032 \text{ when } \mu = 10.1) =$

        $P(-5.08 < Z < .08) = .5319$. Similarly, $\beta(9.8) = P(2.42 < Z < 7.58) = .0078$.

    d. $c = \pm 2.58$

    e. Now $\frac{\sigma}{\sqrt{n}} = \frac{.2}{3.162} = .0632$. Thus 10.1032 is replaced by c, where $\frac{c - 10}{.0632} = 1.96$ i.e. $c = 10.124$.

        Similarly, 9.8968 is replaced by 9.876.

    f. $\overline{x} = 10.020$. Since $\overline{x}$ is neither ≥ 10.124 nor ≤ 9.876, , it is not in the rejection region. H_0 is not rejected; it is still plausible that $\mu = 10$.

    g. $\overline{x} \geq 10.1032$ or ≤ 9.8968 iff $z \geq 2.58$ or $z \leq -2.58$.

13.

    a. $P(\overline{X} \geq \mu_o + 2.33 \frac{\sigma}{\sqrt{n}} \text{ when } \mu = \mu_o) = P\left(Z \geq \dfrac{\left(\mu_o + 2.33 \frac{\sigma}{\sqrt{n}} - \mu_o\right)}{\frac{\sigma}{\sqrt{n}}} \right) = P(Z \geq 2.33) = .01$, where Z is a

    standard normal rv.

    b. $P(\text{reject } H_0 \text{ when } \mu = 99) = P(\overline{X} \geq 102.33 \text{ when } \mu = 99) = P\left(Z \geq \frac{102 - 99}{1} \right) = P(Z \geq 3.33) = .0004$.

        Similarly, $\alpha(98) = P(\overline{X} \geq 102.33 \text{ when } \mu = 98) = P(Z \geq 4.33) \approx 0$. In general, we have $P(\text{type I error})$ $< .01$ when this probability is calculated for a value of μ less than 100. The boundary value, $\mu = 100$, yields the largest α.

Section 8.2

15.

    a. $\alpha = P(Z \geq 1.88 \text{ when } Z \text{ has a standard normal distribution}) = 1 - \Phi(1.88) = .0301$.

    b. $\alpha = P(Z \leq -2.75 \text{ when } Z \text{ has a standard normal distribution}) = \Phi(-2.75) = .0030$.

    c. $\alpha = \Phi(-2.88) + [1 - \Phi(2.88)] = .004$.

17.

a. $z = \dfrac{30{,}960 - 30{,}000}{1500 / \sqrt{16}} = 2.56 > 2.33$ so reject H_0.

b. $\beta(30{,}500) : \Phi\left(2.33 + \dfrac{30{,}000 - 30{,}500}{1500 / \sqrt{16}}\right) = \Phi(1.00) = .8413$.

c. $\beta(30{,}500) = .05 : n = \left[\dfrac{1500(2.33 + 1.645)}{30{,}000 - 30{,}500}\right]^2 = 142.2$, so use $n = 143$.

d. $\alpha = 1 - \Phi(2.56) = .0052$.

19.

a. Reject H_0 if either $z \geq 2.58$ or $z \leq -2.58$; $\dfrac{\sigma}{\sqrt{n}} = 0.3$, so $z = \dfrac{94.32 - 95}{0.3} = -2.27$. Since -2.27 is not in the rejection region, don't reject H_0.

b. $\beta(94) = \Phi\left(2.58 + \dfrac{1}{0.3}\right) - \Phi\left(-2.58 + \dfrac{1}{0.3}\right) = \Phi(5.91) - \Phi(.75) = .2266$.

c. $n = \left[\dfrac{1.20(2.58 + 1.28)}{95 - 94}\right]^2 = 21.46$, so use $n = 22$.

21. With $H_0 : \mu = .5$ and $H_a : \mu \neq .5$ we reject H_0 if $t > t_{\alpha/2, n-1}$ or $t < -t_{\alpha/2, n-1}$.

a. $1.6 < t_{.025, 12} = 2.179$, so don't reject H_0.

b. $-1.6 > -t_{.025, 12} = -2.179$, so don't reject H_0.

c. $-2.6 > -t_{.005, 24} = -2.797$, so don't reject H_0.

d. $-3.9 <$ the negative of all t values in the df $= 24$ row, so we reject H_0 in favor of H_a.

23. $H_0 : \mu = 360$ v. $H_a : \mu > 360$; $t = \dfrac{\bar{x} - 360}{s / \sqrt{n}}$; reject H_0 if $t > t_{.05, 25} = 1.708$; $t = \dfrac{370.69 - 360}{24.36 / \sqrt{26}} = 2.24 > 1.708$.

Thus H_0 should be rejected. There appears to be a contradiction of the prior belief.

25.

a. $H_0 : \mu = 5.5$ v. $H_a : \mu \neq 5.5$; for a level .01 test, (not specified in the problem description), reject H_0 if either $z \geq 2.58$ or $z \leq -2.58$. Since $z = \dfrac{5.25 - 5.5}{.075} = -3.33 \leq -2.58$, reject H_0.

b. $1 - \beta(5.6) = 1 - \Phi\left(2.58 + \dfrac{(-.1)}{.075}\right) + \Phi\left(-2.58 + \dfrac{(-.1)}{.075}\right) = 1 - \Phi(1.25) + \Phi(-3.91) = .105$.

c. $n = \left[\dfrac{.3(2.58 + 2.33)}{-.1}\right]^2 = 216.97$, so use $n = 217$.

27.

a. Using software, $\bar{x} = 0.75$, $\tilde{x} = 0.64$, $s = .3025$, $f_s = 0.48$. These summary statistics, as well as a box plot (not shown) indicate substantial positive skewness, but no outliers.

b. No, it is not plausible from the results in part **a** that the variable ALD is normal. *However*, since $n = 49$, normality is not required for the use of z inference procedures.

c. We wish to test $H_0: \mu \geq 1.0$ versus $H_a: \mu < 1.0$. The test statistic is $z = \dfrac{0.75 - 1.0}{.3025 / \sqrt{49}} = -5.79$; at any reasonable significance level, we reject the null hypothesis. Yes, the data provides strong evidence that the true average ALD is less than 1.0.

d. $\bar{x} + z_{.05} \dfrac{s}{\sqrt{n}} = 0.75 + 1.645 \dfrac{.3025}{\sqrt{49}} = 0.821$

29.

a. The hypotheses are $H_0: \mu = 200$ versus $H_a: \mu > 200$. H_0 will be rejected at level $\alpha = .05$ if $t \geq t_{.05, 12-1} = t_{.05,11} = 1.796$. With the data provided, $t = \dfrac{\bar{x} - \mu_0}{s / \sqrt{n}} = \dfrac{249.7 - 200}{145.1 / \sqrt{12}} = 1.19$. Since $1.19 < 1.796$, H_0 is not rejected at the $\alpha = .05$ level. We have insufficient evidence to conclude that the true average repair time exceeds 200 minutes.

b. With $d = \dfrac{|\mu_0 - \mu|}{\sigma} = \dfrac{|200 - 300|}{150} = 0.67$, df = 11, and $\alpha = .05$, software calculates power $\approx .70$, so $\beta(300) \approx .30$.

31. Define μ = the population mean maximum weight of lift (MAWL) for a frequency of four lifts per minute. The hypotheses are $H_0: \mu = 25$ versus $H_a: \mu > 25$.

We have a random sample from the population of all health males ages 18-30. Although we have little power to detect departures from normality with just $n = 5$, the probability plot below suggests that a normal population distribution is at least somewhat plausible.

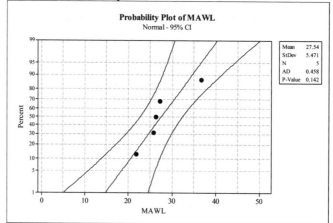

We will reject H_0 if $t \geq t_{.05,4} = 2.132$. With the data provided, $t = \dfrac{\bar{x} - \mu_0}{s / \sqrt{n}} = \dfrac{27.54 - 25}{5.47 / \sqrt{5}} = 1.04$. Since $1.04 < 2.132$, we fail to reject H_0. At the .05 significance level, there is insufficient evidence to conclude that the population mean MAWL, μ, exceeds 25 kg.

112

33. Software provides $\bar{x} = 1.243$ and $s = 0.448$ for this sample.

   a. The parameter of interest is μ = the population mean expense ratio (%) for large-cap growth mutual funds. The hypotheses are $H_0: \mu = 1$ versus $H_a: \mu > 1$.

   We have a random sample, and a normal probability plot is reasonably linear, so the assumptions for a t procedure are met. We'll reject H_0 if $t \geq t_{.01,20-1} = 2.539$.

   The test statistic is $t = \dfrac{1.243 - 1}{0.448 / \sqrt{20}} = 2.43 < 2.539$. Hence, we fail to reject H_0 at the .01 significance

   level. There is insufficient evidence, at the $\alpha = .01$ level, to conclude that the population mean expense ratio for large-cap growth mutual funds exceeds 1%.

   b. A Type I error would be to incorrectly conclude that the population mean expense ratio for large-cap growth mutual funds exceeds 1% when, in fact the mean is 1%. A Type II error would be to fail to recognize that the population mean expense ratio for large-cap growth mutual funds exceeds 1% when that's actually true.

   Since we failed to reject H_0 in (a), we potentially committed a Type II error there. If we later find out that, in fact, $\mu = 1.33$, so H_a was actually true all along, then yes we have committed a Type II error.

   c. With $n = 20$ so df = 19, $d = \dfrac{1.33 - 1}{.5} = .66$, and $\alpha = .01$, software provides power $\approx .66$. (Note: it's purely a coincidence that power and d are the same decimal!) This means that if the true values of μ and σ are $\mu = 1.33$ and $\sigma = .5$, then there is a 66% probability of correctly rejecting $H_0: \mu = 1$ in favor of $H_a: \mu > 1$ at the .01 significance level based upon a sample of size $n = 20$.

35. $\beta(\mu_o - \Delta) = \Phi\!\left(z_{\alpha/2} + \Delta\sqrt{n}/\sigma\right) - \Phi\!\left(-z_{\alpha/2} + \Delta\sqrt{n}/\sigma\right) = 1 - \Phi\!\left(-z_{\alpha/2} - \Delta\sqrt{n}/\sigma\right) - \left[1 - \Phi\!\left(z_{\alpha/2} - \Delta\sqrt{n}/\sigma\right)\right] =$

$\Phi\!\left(z_{\alpha/2} - \Delta\sqrt{n}/\sigma\right) - \Phi\!\left(-z_{\alpha/2} - \Delta\sqrt{n}/\sigma\right) = \beta(\mu_o + \Delta)$

Section 8.3

37.

   a. The parameter of interest is p = the proportion of the population of female workers that have BMIs of at least 30 (and, hence, are obese). The hypotheses are $H_0: p = .20$ versus $H_a: p > .20$.

   With $n = 541$, $np_0 = 541(.2) = 108.2 \geq 10$ and $n(1 - p_0) = 541(.8) = 432.8 \geq 10$, so the "large-sample" z procedure is applicable. Hence, we will reject H_0 if $z \geq z_{.05} = 1.645$.

   From the data provided, $\hat{p} = \dfrac{120}{541} = .2218$, so $z = \dfrac{\hat{p} - p_0}{\sqrt{p_0(1 - p_0)/n}} = \dfrac{.2218 - .20}{\sqrt{.20(.80)/541}} = 1.27$. Since 1.27

   < 1.645, we fail to reject H_0 at the $\alpha = .05$ level. We do not have sufficient evidence to conclude that more than 20% of the population of female workers is obese.

   b. A Type I error would be to incorrectly conclude that more than 20% of the population of female workers is obese, when the true percentage is 20%. A Type II error would be to fail to recognize that more than 20% of the population of female workers is obese when that's actually true.

   c. The question is asking for the chance of committing a Type II error when the true value of p is .25, i.e. $\beta(.25)$. Using the textbook formula,

   $$\beta(.25) = \Phi\!\left[\dfrac{.20 - .25 + 1.645\sqrt{.20(.80)/541}}{\sqrt{.25(.75)/541}}\right] = \Phi(-1.166) \approx .121.$$

113

39.

1 p = true proportion of all donors with type A blood

2 $H_0: p = .40$

3 $H_a: p \neq .40$

4 $z = \dfrac{\hat{p} - p_o}{\sqrt{p_o(1 - p_o)/n}} = \dfrac{\hat{p} - .40}{\sqrt{.40(.60)/n}}$

5 Reject H_0 if $z \geq 2.58$ or $z \leq -2.58$

6 $z = \dfrac{82/150 - .40}{\sqrt{.40(.60)/150}} = \dfrac{.147}{.04} = 3.667$

7 Reject H_0. The data does suggest that the percentage of all donors with type A blood differs from 40%. (at the .01 significance level). Since the z critical value for a significance level of .05 is less than that of .01, the conclusion would not change.

41.

a. The parameter of interest is p = the proportion of all wine customers who would find screw tops acceptable. The hypotheses are $H_0: p = .25$ versus $H_a: p < .25$.
With $n = 106$, $np_0 = 106(.25) = 26.5 \geq 10$ and $n(1 - p_0) = 106(.75) = 79.5 \geq 10$, so the "large-sample" z procedure is applicable. Hence, we will reject H_0 if $z \leq -z_{.10} = -1.28$.

From the data provided, $\hat{p} = \dfrac{22}{106} = .208$, so $z = \dfrac{.208 - .25}{\sqrt{.25(.75)/106}} = -1.01$.

Since $-1.01 > -1.28$, we fail to reject H_0 at the $\alpha = .10$ level. We do not have sufficient evidence to suggest that less than 25% of all customers find screw tops acceptable. Therefore, we recommend that the winery <u>should</u> switch to screw tops.

b. A Type I error would be to incorrectly conclude that less than 25% of all customers find screw tops acceptable, when the true percentage is 25%. Hence, we'd recommend not switching to screw tops when there use is actually justified. A Type II error would be to fail to recognize that less than 25% of all customers find screw tops acceptable when that's actually true. Hence, we'd recommend (as we did in (a)) that the winery switch to screw tops when the switch is not justified. Since we failed to reject H_0 in (a), we may have committed a Type II error.

43.

a. p = true proportion of current customers who qualify. $H_0: p = .05$ v. $H_a: p \neq .05$, $z = \dfrac{\hat{p} - .05}{\sqrt{.05(.95)/n}}$,

reject H_0 if $z \geq 2.58$ or $z \leq -2.58$. $\hat{p} = .08$, so $z = \dfrac{.03}{.00975} = 3.07 \geq 2.58$ and H_0 is rejected. The company's premise is not correct.

b. $\beta(.10) = \Phi\left[\dfrac{.05 - .10 + 2.58\sqrt{.05(.95)/500}}{\sqrt{.10(.90)/500}}\right] - \Phi\left[\dfrac{.05 - .10 - 2.58\sqrt{.05(.95)/500}}{\sqrt{.10(.90)/500}}\right] \approx \Phi(-1.85) - 0 = .0332$

45. The hypotheses are $H_0: p = .10$ v. $H_a: p > .10$, so R has the form $\{c, ..., n\}$.
The values $n = 10$, $c = 3$ (i.e. R = {3, 4, ..., 10}) yield $\alpha = 1 - B(2; 10, .1) = .07$, while no larger R has $\alpha \leq .10$. However, $\beta(.3) = B(2; 10, .3) = .383$.
The values $n = 20$, $c = 5$ yield $\alpha = 1 - B(4; 20, .1) = .043$, but again $\beta(.3) = B(4; 20, .3) = .238$ is too high.
The values $n = 25$, $c = 5$ yield $\alpha = 1 - B(4; 25, .1) = .098$ while $\beta(.7) = B(4; 25, .3) = .090 \leq .10$, so $n = 25$ should be used. The rejection region is $R = \{5,...,25\}$, $\alpha = .098$, and $\beta(.7) = .090$.

114

Section 8.4

47. Using $\alpha = .05$, H_0 should be rejected whenever P-value $< .05$.
   a. P-value $= .001 < .05$, so reject H_0

   b. $.021 < .05$, so reject H_0.

   c. $.078$ is not $< .05$, so don't reject H_0.

   d. $.047 < .05$, so reject H_0 (a close call).

   e. $.148 > .05$, so H_0 can't be rejected at level $.05$.

49. In each case, the P-value equals $P(Z > z) = 1 - \Phi(z)$.
   a. $.0778$

   b. $.1841$

   c. $.0250$

   d. $.0066$

   e. $.5438$

51. Use Table A.8.
   a. $P(t > 2.0)$ at 8df $= .040$.

   b. $P(t < -2.4)$ at 11df $= .018$.

   c. $2P(t < -1.6)$ at 15df $= 2(.065) = .130$.

   d. by symmetry, $P(t > -.4) = 1 - P(t > .4)$ at 19df $= 1 - .347 = .653$.

   e. $P(t > 5.0)$ at 5df $< .005$.

   f. $2P(t < -4.8)$ at 40df $< 2(.000) = .000$ to three decimal places.

53. The P-value is greater than the level of significance $\alpha = .01$, therefore fail to reject H_0. The data does not indicate a statistically significant difference in average serum receptor concentration between pregnant women and all other women.

55. Here we might be concerned with departures above as well as below the specified weight of 5.0, so the relevant hypotheses are $H_0: \mu = 5.0$ v. $H_a: \mu \neq 5.0$. Since $\dfrac{s}{\sqrt{n}} = .035$, $z = \dfrac{-.13}{.035} = -3.71$. Because 3.71 is "off" the z-table, P-value $< 2(.0002) = .0004$, so H_0 should be rejected.

57. The parameter of interest is p = the proportion of all college students who have maintained lifetime abstinence from alcohol. The hypotheses are $H_0: p = .1$, $H_a: p > .1$.

With $n = 462$, $np_0 = 462(.1) = 46.2 \geq 10$ $n(1 - p_0) = 462(.9) = 415.8 \geq 10$, so the "large-sample" z procedure is applicable.

From the data provided, $\hat{p} = \dfrac{51}{462} = .1104$, so $z = \dfrac{.1104 - .1}{\sqrt{.1(.9)/462}} = 0.74$.

The corresponding one-tailed P-value is $P(Z \geq 0.74) = 1 - \Phi(0.74) \approx .23$.

Since $.23 > .05$, we fail to reject H_0 at the $\alpha = .05$ level (and, in fact, at any reasonable significance level). The data does not give evidence to suggest that more than 10% of all college students have completely abstained from alcohol use.

59.

a. The accompanying normal probability plot is acceptably linear, which suggests that a normal population distribution is quite plausible.

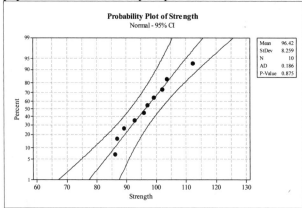

b. The parameter of interest is μ = the true average compression strength (MPa) for this type of concrete. The hypotheses are $H_0: \mu = 100$ versus $H_a: \mu < 100$.

Since the data come from a plausibly normal population, we will use the t procedure. The test statistic is $t = \dfrac{\bar{x} - \mu_0}{s/\sqrt{n}} = \dfrac{96.42 - 100}{8.26/\sqrt{10}} = -1.37$. The corresponding one-tailed P-value, at df = $10 - 1 = 9$, is $P(T \leq -1.37) \approx .102$.

The P-value slightly exceeds .10, the largest α level we'd consider using in practice, so the null hypothesis $H_0: \mu = 100$ should not be rejected. This concrete <u>should</u> be used.

61. μ = true average reading, $H_0: \mu = 70$ v. $H_a: \mu \neq 70$, and $t = \dfrac{\bar{x} - 70}{s/\sqrt{n}} = \dfrac{75.5 - 70}{7/\sqrt{6}} = \dfrac{5.5}{2.86} = 1.92$.

From table A.8, df = 5, P-value = $2[P(t > 1.92)] \approx 2(.058) = .116$. At significance level .05, there is not enough evidence to conclude that the spectrophotometer needs recalibrating.

116

Section 8.5

63.

a. The formula for β is $1 - \Phi\left(-2.33 + \frac{\sqrt{n}}{9}\right)$, which gives .8888 for $n = 100$, .1587 for $n = 900$, and .0006 for $n = 2500$.

b. $Z = -5.3$, which is "off the z table," so P-value $< .0002$; this value of z is quite statistically significant.

c. No. Even when the departure from H_0 is insignificant from a practical point of view, a statistically significant result is highly likely to appear; the test is too likely to detect small departures from H_0.

Supplementary Exercises

65. Because $n = 50$ is large, we use a z test here, rejecting H_0: $\mu = 3.2$ in favor of H_a: $\mu \neq 3.2$ if either $z \geq z_{.025} = 1.96$ or $z \leq -1.96$. The computed z value is $z = \frac{3.05 - 3.20}{.34/\sqrt{50}} = -3.12$. Since -3.12 is ≤ -1.96, H_0 should be rejected in favor of H_a.

67.

a. H_0: $\mu = .85$ v. H_a: $\mu \neq .85$

b. With a P-value of .30, we would reject the null hypothesis at any reasonable significance level, which includes both .05 and .10.

69.

a. The parameter of interest is μ = the true average contamination level (Total Cu, in mg/kg) in this region. The hypotheses are H_0: $\mu = 20$ versus H_a: $\mu > 20$. Using a one-sample t procedure, with $\bar{x} = 45.31$ and SE($\bar{x}$) = 5.26, the test statistic is $t = \frac{45.31 - 20}{5.26} = 3.86$. That's a very large t-statistic; however, at df = $3 - 1 = 2$, the P-value is $P(T \geq 3.86) \approx .03$. (Using the tables with $t = 3.9$ gives a P-value of $\approx .02$.) Since the P-value exceeds .01, we would fail to reject H_0 at the $\alpha = .01$ level. This is quite surprising, given the large t-value (45.31 greatly exceeds 20), but it's a result of the very small n.

b. We want the probability that we fail to reject H_0 in part (a) when $n = 3$ and the true values of μ and σ are $\mu = 50$ and $\sigma = 10$, i.e. $\beta(50)$. Using software, we get $\beta(50) \approx .57$.

71. $n = 47$, $\bar{x} = 215$ mg, $s = 235$ mg, scope of values = 5 mg to 1,176 mg

a.   No, the distribution does not appear to be normal. It appears to be skewed to the right, since 0 is less than one standard deviation below the mean. It is not necessary to assume normality if the sample size is large enough due to the central limit theorem. This sample size is large enough so we can conduct a hypothesis test about the mean.

b.

1	Parameter of interest: μ = true daily caffeine consumption of adult women.
2	H_0: $\mu = 200$
3	H_a: $\mu > 200$
4	$z = \dfrac{\bar{x} - 200}{s / \sqrt{n}}$
5	RR: $z \geq 1.282$ or if P-value $\leq .10$
6	$z = \dfrac{215 - 200}{235 / \sqrt{47}} = .44$; P-value $= 1 - \Phi(.44) = .33$
7	Fail to reject H_0, because $.33 > .10$. The data does not indicate that daily consumption of all adult women exceeds 200 mg.

73.

a.   From Table A.17, when $\mu = 9.5$, $d = .625$, and df = 9, $\beta \approx .60$. When $\mu = 9.0$, $d = 1.25$, and df = 9, $\beta \approx .20$.

b.   From Table A.17, when $\beta = .25$ and $d = .625$, $n \approx 28$.

75.

a.   With H_0: $p = 1/75$ v. H_a: $p \neq 1/75$, we reject H_0 if either $z \geq 1.96$ or $z \leq -1.96$. With $\hat{p} = \dfrac{16}{800} = .02$,

$z \dfrac{.02 - .01333}{\sqrt{\dfrac{.01333(.98667)}{800}}} = 1.645$, which is not in either rejection region. Thus, we fail to reject the null

hypothesis. There is not evidence that the incidence rate among prisoners differs from that of the adult population. The possible error we could have made is a type II.

b. P–value $= 2[1 - \Phi(1.645)] = 2[.05] = .10$. Yes, since $.10 < .20$, we could reject H_0.

77.   Even though the underlying distribution may not be normal, a z test can be used because n is large. The null hypothesis H_0: $\mu = 3200$ should be rejected in favor of H_a: $\mu < 3200$ if $z \leq -z_{.001} = -3.08$. The computed test statistic is $z = \dfrac{3107 - 3200}{188 / \sqrt{45}} = -3.32 \leq -3.08$, so H_0 should be rejected at level .001.

79.   We wish to test H_0: $\mu = 4$ versus H_a: $\mu > 4$ using the test statistic $z = \dfrac{\bar{x} - 4}{\sqrt{4 / n}}$. For the given sample, $n = 36$

and $\bar{x} = \dfrac{160}{36} = 4.444$, so $z = \dfrac{4.444 - 4}{\sqrt{4 / 36}} = 1.33$.

At level .02, we reject H_0 if $z \geq z_{.02} \approx 2.05$ (since $1 - \Phi(2.05) = .0202$). Because $1.33 < 2.05$, H_0 should not be rejected at this level. We do not have significant evidence at the .02 level to conclude that the true mean of this Poisson process is greater than 4.

81. Let μ = the true mean time, in minutes, for this brand of hot tub to reach 100°F. The hypotheses are H_0: μ = 15 versus H_a: $\mu > 15$. Because the sample size is greater than 40, we'll rely on the robustness of the one-sample t procedure even though the normality assumption might not be satisfied.

From the data provided, the test statistic is $t = \dfrac{\bar{x}-15}{s/\sqrt{n}} = \dfrac{16.5-15}{2.2/\sqrt{42}} = 4.42$. This t statistic is "off the chart" of

Table A.8, so P-value $\approx 0 < .05$, and H_0 is rejected in favor of the conclusion that the true average time exceeds 15 minutes.

83. The 20 df row of Table A.7 shows that $\chi^2_{.99,20} = 8.26 < 8.58$ (H_0 not rejected at level .01) and

$8.58 < 9.591 = \chi^2_{.975,20}$ (H_0 rejected at level .025). Thus $.01 < P$-value $< .025$ and H_0 cannot be rejected at

level .01 (the P-value is the smallest alpha at which rejection can take place, and this exceeds .01).

85.

a. When H_0 is true, $2\lambda_0\Sigma X_i = 2\sum \dfrac{X_i}{\mu_0}$ has a chi-squared distribution with df = $2n$. If the alternative is

H_a: $\mu > \mu_0$, large test statistic values (large Σx_i, since $\bar{x}$ is large) suggest that H_0 be rejected in favor of

H_a, so rejecting when $2\sum \dfrac{X_i}{\mu_0} \geq \chi^2_{\alpha,2n}$ gives a test with significance level α. If the alternative is

H_a: $\mu < \mu_0$, rejecting when $2\sum \dfrac{X_i}{\mu_0} \leq \chi^2_{1-\alpha,2n}$ gives a level α test. The rejection region for H_a: $\mu \neq \mu_0$ is

$2\sum \dfrac{X_i}{\mu_0} \geq \chi^2_{\alpha/2,2n}$ or $2\sum \dfrac{X_i}{\mu_0} \leq \chi^2_{1-\alpha/2,2n}$.

b. H_0: $\mu = 75$ v. H_a: $\mu < 75$. The test statistic value is $\dfrac{2(737)}{75} = 19.65$. At level .01, H_0 is rejected if

$2\sum \dfrac{X_i}{\mu_0} \leq \chi^2_{.99,20} = 8.260$. Clearly 19.65 is not in the rejection region, so H_0 should not be rejected.

The sample data does not suggest that true average lifetime is less than the previously claimed value.

87.

a. $\alpha = P(X \leq 5 \text{ when } p = .9) = B(5; 10, .9) = .002$, so the region (0, 1, …, 5) does specify a level .01 test.

b. The first value to be placed in the upper-tailed part of a two tailed region would be 10, but $P(X = 10 \text{ when } p = .9) = .349$, so whenever 10 is in the rejection region, $\alpha \geq .349$.

c. $\beta(p') = P(X \text{ is } \underline{not} \text{ in } R \text{ when } p = p') = P(X > 5 \text{ when } p = p') = 1 - B(5;10,p')$. The test has no ability to detect a false null hypothesis when $p > .90$ (see the graph for $.90 < p' < 1$). This is a by-product of the unavoidable one-sided rejection region (see **a** and **b**). The test also has an undesirably high $\beta(p')$ for medium-to-large p', a result of the small sample size.

A graph of $\beta(p')$ versus p' appears on the next page.

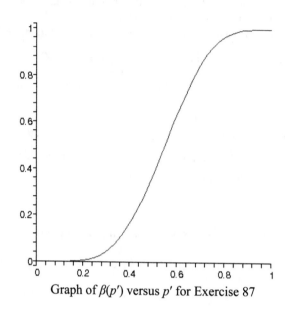

Graph of $\beta(p')$ versus p' for Exercise 87

CHAPTER 9

Section 9.1

1.

    a. $E(\bar{X} - \bar{Y}) = E(\bar{X}) - E(\bar{Y}) = 4.1 - 4.5 = -.4$, irrespective of sample sizes.

    b. $V(\bar{X} - \bar{Y}) = V(\bar{X}) + V(\bar{Y}) = \dfrac{\sigma_1^2}{m} + \dfrac{\sigma_2^2}{n} = \dfrac{(1.8)^2}{100} + \dfrac{(2.0)^2}{100} = .0724$, and the SD of $\bar{X} - \bar{Y}$ is

        $\bar{X} - \bar{Y} = \sqrt{.0724} = .2691$.

    c. A normal curve with mean and s.d. as given in **a** and **b** (because $m = n = 100$, the CLT implies that both $\bar{X}$ and $\bar{Y}$ have approximately normal distributions, so $\bar{X} - \bar{Y}$ does also). The shape is not necessarily that of a normal curve when $m = n = 10$, because the CLT cannot be invoked. So if the two lifetime population distributions are not normal, the distribution of $\bar{X} - \bar{Y}$ will typically be quite complicated.

3.   The test statistic value is $z = \dfrac{(\bar{x} - \bar{y}) - 5000}{\sqrt{\dfrac{s_1^2}{m} + \dfrac{s_2^2}{n}}}$, and H_0 will be rejected at level .01 if $z \geq 2.33$. We compute

$z = \dfrac{(42{,}500 - 36{,}800) - 5000}{\sqrt{\dfrac{2200^2}{45} + \dfrac{1500^2}{45}}} = \dfrac{700}{396.93} = 1.76$, which is not ≥ 2.33, so we don't reject H_0 and conclude that

the true average life for radials does not exceed that for economy brand by significantly more than 500.

5.

    a. H_a says that the average calorie output for sufferers is more than 1 cal/cm^2/min below that for non-sufferers. $\sqrt{\dfrac{\sigma_1^2}{m} + \dfrac{\sigma_2^2}{n}} = \sqrt{\dfrac{(.2)^2}{10} + \dfrac{(.4)^2}{10}} = .1414$, so $z = \dfrac{(.64 - 2.05) - (-1)}{.1414} = -2.90$. At level .01, H_0 is rejected if $z \leq -2.33$; since $-2.90 \leq -2.33$, reject H_0.

    b. From **a**, P-value $= \Phi(-2.90) = .0019$.

    c. $\beta = 1 - \Phi\left(-2.33 - \dfrac{-1.2 + 1}{.1414}\right) = 1 - \Phi(-.92) = .8212$.

    d. $m = n = \dfrac{.2(2.33 + 1.28)^2}{(-.2)^2} = 65.15$, so use 66.

121

7. Let μ_1 denote the true mean course GPA for all courses taught by full-time faculty, and let μ_2 denote the true mean course GPA for all courses taught by part-time faculty. The hypotheses of interest are $H_0: \mu_1 = \mu_2$ versus $H_a: \mu_1 \neq \mu_2$; or, equivalently, $H_0: \mu_1 - \mu_2 = 0$ v. $H_a: \mu_1 - \mu_2 \neq 0$.

The large-sample test statistic is $z = \dfrac{(\bar{x} - \bar{y}) - \Delta_0}{\sqrt{\dfrac{s_1^2}{m} + \dfrac{s_2^2}{n}}} = \dfrac{(2.7186 - 2.8639) - 0}{\sqrt{\dfrac{(.63342)^2}{125} + \dfrac{(.49241)^2}{88}}} = -1.88$. The corresponding

two-tailed P-value is $P(|Z| \geq |-1.88|) = 2[1 - \Phi(1.88)] = .0602$.

Since the P-value exceeds $\alpha = .01$, we fail to reject H_0. At the .01 significance level, there is insufficient evidence to conclude that the true mean course GPAs differ for these two populations of faculty.

9.

   a. Point estimate $\bar{x} - \bar{y} = 19.9 - 13.7 = 6.2$. It appears that there could be a difference.

   b. $H_0: \mu_1 - \mu_2 = 0$, $H_a: \mu_1 - \mu_2 \neq 0$, $z = \dfrac{(19.9 - 13.7)}{\sqrt{\dfrac{39.1^2}{60} + \dfrac{15.8^2}{60}}} = \dfrac{6.2}{5.44} = 1.14$, and the P-value = $2[P(Z > 1.14)] =$

      $2(.1271) = .2542$. The P-value is larger than any reasonable α, so we do not reject H_0. There is no statistically significant difference.

   c. No. With a normal distribution, we would expect most of the data to be within 2 standard deviations of the mean, and the distribution should be symmetric. Two sd's above the mean is 98.1, but the distribution stops at zero on the left. The distribution is positively skewed.

   d. We will calculate a 95% confidence interval for μ, the true average length of stays for patients given the treatment. $19.9 \pm 1.96 \dfrac{39.1}{\sqrt{60}} = 19.9 \pm 9.9 = (10.0, 21.8)$.

11. $(\bar{x} - \bar{y}) \pm z_{\alpha/2} \sqrt{\dfrac{s_1^2}{m} + \dfrac{s_2^2}{n}} = (\bar{x} - \bar{y}) \pm z_{\alpha/2} \sqrt{(SE_1)^2 + (SE_2)^2}$. Using $\alpha = .05$ and $z_{\alpha/2} = 1.96$ yields

$(5.5 - 3.8) \pm 1.96 \sqrt{(0.3)^2 + (0.2)^2} = (0.99, 2.41)$. We are 95% confident that the true average blood lead level for male workers is between 0.99 and 2.41 higher than the corresponding average for female workers.

13. $\sigma_1 = \sigma_2 = .05$, $d = .04$, $\alpha = .01$, $\beta = .05$, and the test is one-tailed $\Rightarrow$

$n = \dfrac{(.0025 + .0025)(2.33 + 1.645)^2}{.0016} = 49.38$, so use $n = 50$.

15.

   a. As either m or n increases, SD decreases, so $\dfrac{\mu_1 - \mu_2 - \Delta_0}{SD}$ increases (the numerator is positive), so

      $\left(z_\alpha - \dfrac{\mu_1 - \mu_2 - \Delta_0}{SD} \right)$ decreases, so $\beta = \Phi \left(z_\alpha - \dfrac{\mu_1 - \mu_2 - \Delta_0}{SD} \right)$ decreases.

   b. As β decreases, z_β increases, and since z_β is the numerator of n, n increases also.

Section 9.2

17.

a. $\quad v = \dfrac{\left(\frac{5^2}{10} + \frac{6^2}{10}\right)^2}{\dfrac{\left(\frac{5^2}{10}\right)^2}{9} + \dfrac{\left(\frac{6^2}{10}\right)^2}{9}} = \dfrac{37.21}{.694 + 1.44} = 17.43 \approx 17.$

b. $\quad v = \dfrac{\left(\frac{5^2}{10} + \frac{6^2}{15}\right)^2}{\dfrac{\left(\frac{5^2}{10}\right)^2}{9} + \dfrac{\left(\frac{6^2}{15}\right)^2}{14}} = \dfrac{24.01}{.694 + .411} = 21.7 \approx 21.$

c. $\quad v = \dfrac{\left(\frac{2^2}{10} + \frac{6^2}{15}\right)^2}{\dfrac{\left(\frac{2^2}{10}\right)^2}{9} + \dfrac{\left(\frac{6^2}{15}\right)^2}{14}} = \dfrac{7.84}{.018 + .411} = 18.27 \approx 18.$

d. $\quad v = \dfrac{\left(\frac{5^2}{12} + \frac{6^2}{24}\right)^2}{\dfrac{\left(\frac{5^2}{12}\right)^2}{11} + \dfrac{\left(\frac{6^2}{24}\right)^2}{23}} = \dfrac{12.84}{.395 + .098} = 26.05 \approx 26.$

19. For the given hypotheses, the test statistic is $t = \dfrac{115.7 - 129.3 + 10}{\sqrt{\frac{5.03^2}{6} + \frac{5.38^2}{6}}} = \dfrac{-3.6}{3.007} = -1.20$, and the df is

$v = \dfrac{\left(4.2168 + 4.8241\right)^2}{\dfrac{\left(4.2168\right)^2}{5} + \dfrac{\left(4.8241\right)^2}{5}} = 9.96$, so use df = 9. We will reject H_0 if $t \le -t_{.01,9} = -2.764$;

since $-1.20 > -2.764$, we don't reject H_0.

21. Let $\mu_1 =$ the true average gap detection threshold for normal subjects, and $\mu_2 =$ the corresponding value for CTS subjects. The relevant hypotheses are $H_0: \mu_1 - \mu_2 = 0$ v. $H_a: \mu_1 - \mu_2 < 0$, and the test statistic is

$t = \dfrac{1.71 - 2.53}{\sqrt{.0351125 + .07569}} = \dfrac{-.82}{.3329} = -2.46$. Using df $v = \dfrac{\left(.0351125 + .07569\right)^2}{\dfrac{\left(.0351125\right)^2}{7} + \dfrac{\left(.07569\right)^2}{9}} = 15.1$, or 15, the

rejection region is $t \le -t_{.01,15} = -2.602$. Since -2.46 is not ≤ -2.602, we fail to reject H_0. We have insufficient evidence to claim that the true average gap detection threshold for CTS subjects exceeds that for normal subjects.

23.

    a. Using Minitab to generate normal probability plots, we see that both plots illustrate sufficient linearity. Therefore, it is plausible that both samples have been selected from normal population distributions.

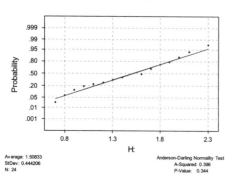

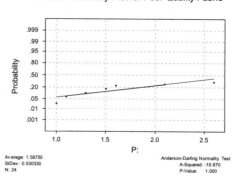

    b. The comparative boxplot does not suggest a difference between average extensibility for the two types of fabrics.

Comparative Box Plot for High Quality and Poor Quality Fabric

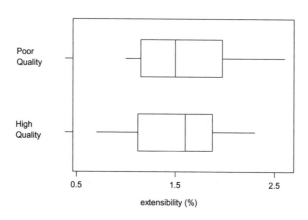

    c. We test $H_0 : \mu_1 - \mu_2 = 0$ v. $H_a : \mu_1 - \mu_2 \neq 0$. With degrees of freedom $v = \dfrac{(.0433265)^2}{.00017906} = 10.5$, which we round down to 10, and using significance level .05 (not specified in the problem), we reject H_0 if $|t| \geq t_{.025,10} = 2.228$. The test statistic is $t = \dfrac{-.08}{\sqrt{(.0433265)}} = -.38$, which is not ≥ 2.228 in absolute value, so we cannot reject H_0. There is insufficient evidence to claim that the true average extensibility differs for the two types of fabrics.

25. We calculate the degrees of freedom $v = \dfrac{\left(\frac{5.5^2}{28} + \frac{7.8^2}{31}\right)^2}{\dfrac{\left(\frac{5.5^2}{28}\right)^2}{27} + \dfrac{\left(\frac{7.8^2}{31}\right)^2}{30}} = 53.95$, or about 54 (normally we would round

down to 53, but this number is very close to 54—of course for this large number of df, using either 53 or 54 won't make much difference in the critical t value) so the desired confidence interval is

$(91.5 - 88.3) \pm 1.68\sqrt{\frac{5.5^2}{28} + \frac{7.8^2}{31}} = 3.2 \pm 2.931 = (.269, 6.131)$. Because 0 does not lie inside this interval, we

can be reasonably certain that the true difference $\mu_1 - \mu_2$ is not 0 and, therefore, that the two population means are not equal. For a 95% interval, the t value increases to about 2.01 or so, which results in the interval 3.2 ± 3.506. Since this interval does contain 0, we can no longer conclude that the means are different if we use a 95% confidence interval.

27.

   a. Let's construct a 99% CI for μ_{AN}, the true mean intermuscular adipose tissue (IAT) under the described AN protocol. Assuming the data comes from a normal population, the CI is given by

$\bar{x} \pm t_{\alpha/2,n-1}\dfrac{s}{\sqrt{n}} = .52 \pm t_{.005,15}\dfrac{.26}{\sqrt{16}} = .52 \pm 2.947\dfrac{.26}{\sqrt{16}} = (.33, .71)$. We are 99% confident that the true mean

IAT under the AN protocol is between .33 kg and .71 kg.

   b. Let's construct a 99% CI for $\mu_{AN} - \mu_C$, the difference between true mean AN IAT and true mean control IAT. Assuming the data come from normal populations, the CI is given by

$(\bar{x} - \bar{y}) \pm t_{\alpha/2,v}\sqrt{\dfrac{s_1^2}{m} + \dfrac{s_2^2}{n}} = (.52 - .35) \pm t_{.005,21}\sqrt{\dfrac{(.26)^2}{16} + \dfrac{(.15)^2}{8}} = .17 \pm 2.831\sqrt{\dfrac{(.26)^2}{16} + \dfrac{(.15)^2}{8}} = (-.07, .41)$.

Since this CI includes zero, it's plausible that the difference between the two true means is zero (i.e., $\mu_{AN} - \mu_C = 0$). [Note: the df calculation $v = 21$ comes from applying the formula in the textbook.]

29. Let μ_1 = the true average compression strength for strawberry drink and let μ_2 = the true average compression strength for cola. A lower tailed test is appropriate. We test $H_0: \mu_1 - \mu_2 = 0$ v. $H_a: \mu_1 - \mu_2 < 0$.

The test statistic is $t = \dfrac{-14}{\sqrt{29.4 + 15}} = -2.10$; $v = \dfrac{(44.4)^2}{\dfrac{(29.4)^2}{14} + \dfrac{(15)^2}{14}} = \dfrac{1971.36}{77.8114} = 25.3$, so use df=25.

The P-value $\approx P(t < -2.10) = .023$. This P-value indicates strong support for the alternative hypothesis. The data does suggest that the extra carbonation of cola results in a higher average compression strength.

31.

a. The most notable feature of these boxplots is the larger amount of variation present in the mid-range data compared to the high-range data. Otherwise, both look reasonably symmetric with no outliers present.

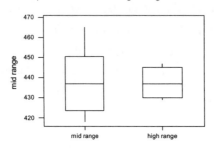

Comparative Box Plot for High Range and Mid Range

b. Using df = 23, a 95% confidence interval for $\mu_{\text{mid-range}} - \mu_{\text{high-range}}$ is

$$\left(438.3 - 437.45\right) \pm 2.069\sqrt{\tfrac{15.1^2}{17} + \tfrac{6.83^2}{11}} = .85 \pm 8.69 = \left(-7.84, 9.54\right).$$ Since plausible values for

$\mu_{\text{mid-range}} - \mu_{\text{high-range}}$ are both positive and negative (i.e., the interval spans zero) we would conclude that there is not sufficient evidence to suggest that the average value for mid-range and the average value for high-range differ.

33. Let μ_1 and μ_2 represent the true mean body mass decrease for the vegan diet and the control diet, respectively. We wish to test the hypotheses $H_0: \mu_1 - \mu_2 \leq 1$ v. $H_a: \mu_1 - \mu_2 > 1$. The relevant test statistic is

$$t = \frac{(5.8 - 3.8) - 1}{\sqrt{\dfrac{3.2^2}{32} + \dfrac{2.8^2}{32}}} = 1.33,$$ with estimated df = 60 using the formula. Rounding to $t = 1.3$, Table A.8 gives a

one-sided *P*-value of .098 (a computer will give the more accurate *P*-value of .094).
Since our *P*-value > α = .05, we fail to reject H_0 at the 5% level. We do not have statistically significant evidence that the true average weight loss for the vegan diet exceeds that for the control diet by more than 1 kg.

35. There are two changes that must be made to the procedure we currently use. First, the equation used to

compute the value of the *t* test statistic is: $t = \dfrac{(\bar{x} - \bar{y}) - \Delta}{s_p\sqrt{\dfrac{1}{m} + \dfrac{1}{n}}}$ where s_p is defined as in Exercise 34 above.

Second, the degrees of freedom = m + n – 2. Assuming equal variances in the situation from Exercise 33,

we calculate s_p as follows: $s_p = \sqrt{\left(\dfrac{7}{16}\right)(2.6)^2 + \left(\dfrac{9}{16}\right)(2.5)^2} = 2.544$. The value of the test statistic is,

then, $t = \dfrac{(32.8 - 40.5) - (-5)}{2.544\sqrt{\dfrac{1}{8} + \dfrac{1}{10}}} = -2.24 \approx -2.2$ with df = 16, and the *P*-value is $P(T < -2.2) = .021$. Since

.021 > .01, we fail to reject H_0.

Section 9.3

37.

a. This exercise calls for paired analysis. First, compute the difference between indoor and outdoor concentrations of hexavalent chromium for each of the 33 houses. These 33 differences are summarized as follows: $n = 33$, $\bar{d} = -.4239$, $s_D = .3868$, where d = (indoor value – outdoor value). Then $t_{.025,32} = 2.037$, and a 95% confidence interval for the population mean difference between indoor and outdoor concentration is $-.4239 \pm (2.037)\left(\dfrac{.3868}{\sqrt{33}}\right) = -.4239 \pm .13715 = (-.5611, -.2868)$. We can be highly confident, at the 95% confidence level, that the true average concentration of hexavalent chromium outdoors exceeds the true average concentration indoors by between .2868 and .5611 nanograms/m^3.

b. A 95% prediction interval for the difference in concentration for the 34th house is $\bar{d} \pm t_{.025,32}\left(s_D\sqrt{1+\tfrac{1}{n}}\right) = -.4239 \pm (2.037)\left(.3868\sqrt{1+\tfrac{1}{33}}\right) = (-1.224, .3758)$. This prediction interval means that the indoor concentration may exceed the outdoor concentration by as much as .3758 nanograms/m^3 and that the outdoor concentration may exceed the indoor concentration by a much as 1.224 nanograms/m^3, for the 34th house. Clearly, this is a wide prediction interval, largely because of the amount of variation in the differences.

39.

a. The accompanying normal probability plot shows that the <u>differences</u> are consistent with a normal population distribution.

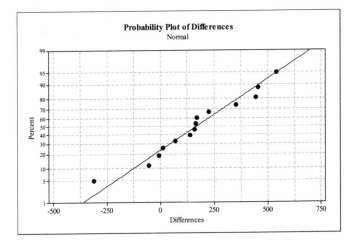

b. We want to test $H_0: \mu_D = 0$ versus $H_a: \mu_D \neq 0$. The test statistic is $t = \dfrac{\bar{d} - 0}{s_D/\sqrt{n}} = \dfrac{167.2 - 0}{228/\sqrt{14}} = 2.74$, and the two-tailed P-value is given by $2[P(T > 2.74)] \approx 2[P(T > 2.7)] = 2[.009] = .018$. Since $.018 < .05$, we reject H_0. There is evidence to support the claim that the true average difference between intake values measured by the two methods is <u>not</u> 0.

41.

a. Let μ_D denote the true mean change in total cholesterol under the aripiprazole regimen. A 95% CI for μ_D, using the "large-sample" method, is $\bar{d} \pm z_{\alpha/2} \dfrac{s_D}{\sqrt{n}} = 3.75 \pm 1.96(3.878) = (-3.85, 11.35)$.

b. Now let μ_D denote the true mean change in total cholesterol under the quetiapine regimen. The hypotheses are $H_0: \mu_D = 0$ versus $H_a: \mu_D > 0$. Assuming the distribution of cholesterol changes under this regimen is normal, we may apply a paired t test:

$$t = \frac{\bar{d} - \Delta_0}{s_D / \sqrt{n}} = \frac{9.05 - 0}{4.256} = 2.126 \Rightarrow P\text{-value} = P(T_{35} \geq 2.126) \approx P(T_{35} \geq 2.1) = .02.$$

Our conclusion depends on our significance level. At the $\alpha = .05$ level, there is evidence that the true mean change in total cholesterol under the quetiapine regimen is positive (i.e., there's been an increase); however, we do not have sufficient evidence to draw that conclusion at the $\alpha = .01$ level.

c. Using the "large-sample" procedure again, the 95% CI is $\bar{d} \pm 1.96 \dfrac{s_D}{\sqrt{n}} = \bar{d} \pm 1.96 SE(\bar{d})$. If this equals $(7.38, 9.69)$, then midpoint $= \bar{d} = 8.535$ and width $= 2(1.96\, SE(\bar{d})) = 9.69 - 7.38 = 2.31 \Rightarrow$ $SE(\bar{d}) = \dfrac{2.31}{2(1.96)} = .59$. Now, use these values to construct a 99% CI (again, using a "large-sample" z method): $\bar{d} \pm 2.576 SE(\bar{d}) = 8.535 \pm 2.576(.59) = 8.535 \pm 1.52 = (7.02, 10.06)$.

43.

a. No. The statement $\mu_D = 0$ implies that, on the average, the time difference between onset of symptoms and diagnosis of Cushing's disease is 0 months! That's impossible, since doctors wouldn't run the tests to detect Cushing's disease until <u>after</u> a child has shown symptoms of the disease (Cushing screening is not a standard preventive procedure). For each child, the difference $d = $ (age at onset) $-$ (age at diagnosis) must be negative.

b. Using the subtraction order in (a), which matches the data in Exercise 43, we wish to test the hypotheses $H_0: \mu_D = -25$ versus $H_a: \mu_D < -25$ (this corresponds to age at diagnosis exceeding age at onset by more than 25 months, on average). The paired t statistic is $t = \dfrac{\bar{d} - \Delta_0}{s_D / \sqrt{n}} = \dfrac{-38.60 - (-25)}{23.18 / \sqrt{15}} = -$ 2.27, and the one-tailed P-value is $P(T_{14} \leq -2.27) = P(T_{14} \geq 2.27) \approx P(T_{14} \geq 2.3) = .019$. This is a low P-value, so we have reasonably compelling evidence that, on the average, the first diagnosis of Cushing's disease happens more than 25 months after the first onset of symptoms.

45.

    a.   Yes, it's quite plausible that the population distribution of differences is normal, since the accompanying normal probability plot of the differences is quite linear.

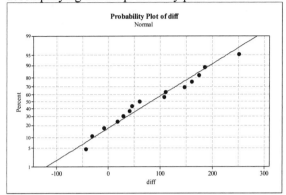

    b.   No. Since the data is paired, the sample means and standard deviations are not useful summaries for inference. Those statistics would only be useful if we were analyzing two <u>independent</u> samples of data. (We could deduce $\bar{d}$ by subtracting the sample means, but there's no way we could deduce s_D from the separate sample standard deviations.)

    c.   The hypotheses corresponding to an upper-tailed test are $H_0: \mu_D = 0$ versus $H_a: \mu_D > 0$. From the data provided, the paired t test statistic is $t = \dfrac{\bar{d} - \Delta_0}{s_D / \sqrt{n}} = \dfrac{82.5 - 0}{87.4 / \sqrt{15}} = 3.66$. The corresponding P-value is $P(T_{14} \geq 3.66) \approx P(T_{14} \geq 3.7) = .001$. While the P-value stated in the article is inaccurate, the conclusion remains the same: we have strong evidence to suggest that the mean difference in ER velocity and IR velocity is positive. Since the measurements were <u>negative</u> (e.g. –130.6 deg/sec and –98.9 deg/sec), this actually means that the magnitude of IR velocity is significantly higher, on average, than the magnitude of ER velocity, as the authors of the article concluded.

47.   From the data, $n = 12$, $\bar{d} = -0.73$, $s_D = 2.81$.

    a.   Let μ_D = the true mean difference in strength between curing under moist conditions and laboratory drying conditions. A 95% CI for μ_D is $\bar{d} \pm t_{.025,11} s_D / \sqrt{n} = -0.73 \pm 2.201(2.81)/\sqrt{10} =$ (–2.52 MPa, 1.05 MPa). In particular, this interval estimate includes the value zero, suggesting that true mean strength is not significantly different under these two conditions.

    b.   Since $n = 12$, we must check that the <u>differences</u> are plausibly from a normal population. The normal probability plot below strongly substantiates that condition.

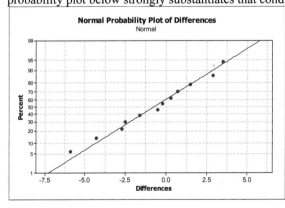

Section 9.4

49. H_0 will be rejected if $z \le -z_{.01} = -2.33$. With $\hat{p}_1 = .150$, and $\hat{p}_2 = .300$, $\hat{p} = \dfrac{30+80}{200+600} = \dfrac{210}{800} = .263$, and

$\hat{q} = .737$. The calculated test statistic is $z = \dfrac{.150-.300}{\sqrt{(.263)(.737)\left(\frac{1}{200}+\frac{1}{600}\right)}} = \dfrac{-.150}{.0359} = -4.18$. Because

$-4.18 \le -2.33$, H_0 is rejected; the proportion of those who repeat after inducement appears lower than those who repeat after no inducement.

51.

1   Parameter of interest: $p_1 - p_2$ = true difference in proportions of those responding to two different survey covers (1 = Plain, 2 = Picture).

2     $H_0 : p_1 - p_2 = 0$

3     $H_a : p_1 - p_2 < 0$

4     $z = \dfrac{\hat{p}_1 - \hat{p}_2}{\sqrt{\hat{p}\hat{q}\left(\frac{1}{m}+\frac{1}{n}\right)}}$

5   Reject H_0 if P-value $< .10$

6     $z = \dfrac{\frac{104}{207} - \frac{109}{213}}{\sqrt{\left(\frac{213}{420}\right)\left(\frac{207}{420}\right)\left(\frac{1}{207}+\frac{1}{213}\right)}} = -.1910$; P-value = .4247

7   Fail to Reject H_0. The data does not indicate that plain cover surveys have a lower response rate.

53.

a. Let p_1 and p_2 denote the true incidence rates of GI problems for the olestra and control groups, respectively. We wish to test $H_0: p_1 - \mu_2 = 0$ v. $H_a: p_1 - p_2 \ne 0$. The pooled proportion is

$\hat{p} = \dfrac{529(.176)+563(.158)}{529+563} = .1667$, from which the relevant test statistic is $z =$

$\dfrac{.176-.158}{\sqrt{(.1667)(.8333)[529^{-1}+563^{-1}]}} = 0.78$. The two-sided P-value is $2P(Z \ge 0.78) = .433 > \alpha = .05$,

hence we fail to reject the null hypothesis. The data do not suggest a statistically significant difference between the incidence rates of GI problems between the two groups.

b. $n = \dfrac{\left(1.96\sqrt{(.35)(1.65)/2}+1.28\sqrt{(.15)(.85)+(.2)(.8)}\right)^2}{(.05)^2} = 1210.39$, so a common sample size of $m = n = $

1211 would be required.

55.

a. A 95% large sample confidence interval formula for $\ln(\theta)$ is $\ln(\hat{\theta}) \pm z_{\alpha/2}\sqrt{\dfrac{m-x}{mx}+\dfrac{n-y}{ny}}$. Taking the

antilogs of the upper and lower bounds gives the confidence interval for θ itself.

b. $\hat{\theta} = \dfrac{\frac{189}{11,034}}{\frac{104}{11,037}} = 1.818$, $\ln(\hat{\theta}) = .598$, and the standard deviation is

$\sqrt{\dfrac{10,845}{(11,034)(189)}+\dfrac{10,933}{(11,037)(104)}} = .1213$, so the CI for $\ln(\theta)$ is $.598 \pm 1.96(.1213) = (.360,.836)$.

Then taking the antilogs of the two bounds gives the CI for θ to be $(1.43, 2.31)$. We are 95% confident that people who do not take the aspirin treatment are between 1.43 and 2.31 times more likely to suffer a heart attack than those who do. This suggests aspirin therapy may be effective in reducing the risk of a heart attack.

57. $\hat{p}_1 = \dfrac{15+7}{40} = .550$, $\hat{p}_2 = \dfrac{29}{42} = .690$, and the 95% CI is $(.550 - .690) \pm 1.96(.106) = -.14 \pm .21 = (-.35, .07)$.

Section 9.5

59.

   a. From Table A.9, column 5, row 8, $F_{.01,5,8} = 3.69$.

   b. From column 8, row 5, $F_{.01,8,5} = 4.82$.

   c. $F_{.95,5,8} = \dfrac{1}{F_{.05,8,5}} = .207$.

   d. $F_{.95,8,5} = \dfrac{1}{F_{.05,5,8}} = .271$

   e. $F_{.01,10,12} = 4.30$

   f. $F_{.99,10,12} = \dfrac{1}{F_{.01,12,10}} = \dfrac{1}{4.71} = .212$.

   g. $F_{.05,6,4} = 6.16$, so $P(F \le 6.16) = .95$.

   h. Since $F_{.99,10,5} = \dfrac{1}{5.64} = .177$, $P(.177 \le F \le 4.74) = P(F \le 4.74) - P(F \le .177) = .95 - .01 = .94$.

61. We test $H_0 : \sigma_1^2 = \sigma_2^2$ v. $H_a : \sigma_1^2 \ne \sigma_2^2$. The calculated test statistic is $f = \dfrac{(2.75)^2}{(4.44)^2} = .384$. With numerator df $= m - 1 = 10 - 1 = 9$ and denominator df $= n - 1 = 5 - 1 = 4$, we reject H_0 if $f \ge F_{.05,9,4} = 6.00$ or $f \le F_{.95,9,4} = \dfrac{1}{F_{.05,4,9}} = \dfrac{1}{3.63} = .275$. Since .384 is in neither rejection region, we do not reject H_0 and conclude that there is no significant difference between the two standard deviations.

63. Let $\sigma_1^2 =$ variance in weight gain for low-dose treatment, and $\sigma_2^2 =$ variance in weight gain for control condition. We wish to test $H_0 : \sigma_1^2 = \sigma_2^2$ v. $H_a : \sigma_1^2 > \sigma_2^2$. The test statistic is $f = \dfrac{s_1^2}{s_2^2}$, and we reject H_0 at level .05 if $f > F_{.05,19,22} \approx 2.08$. $f = \dfrac{(54)^2}{(32)^2} = 2.85 \ge 2.08$, so reject H_0 at level .05. The data does suggest that there is more variability in the low-dose weight gains.

65. $P\left(F_{1-\alpha/2,m-1,n-1} \leq \dfrac{S_1^2/\sigma_1^2}{S_2^2/\sigma_2^2} \leq F_{\alpha/2,m-1,n-1} \right) = 1-\alpha$. The set of inequalities inside the parentheses is clearly

equivalent to $\dfrac{S_2^2 F_{1-\alpha/2,m-1,n-1}}{S_1^2} \leq \dfrac{\sigma_2^2}{\sigma_1^2} \leq \dfrac{S_2^2 F_{\alpha/2,m-1,n-1}}{S_1^2}$. Substituting the sample values s_1^2 and s_2^2 yields the

confidence interval for $\dfrac{\sigma_2^2}{\sigma_1^2}$, and taking the square root of each endpoint yields the confidence interval for

$\dfrac{\sigma_2}{\sigma_1}$. $m = n = 4$, so we need $F_{.05,3,3} = 9.28$ and $F_{.95,3,3} = \dfrac{1}{9.28} = .108$. Then with $s_1 = .160$ and $s_2 = .074$, the

CI for $\dfrac{\sigma_2^2}{\sigma_1^2}$ is (.023, 1.99), and for $\dfrac{\sigma_2}{\sigma_1}$ is (.15, 1.41).

Supplementary Exercises

67. We test $H_0 : \mu_1 - \mu_2 = 0$ v. $H_a : \mu_1 - \mu_2 \neq 0$. The test statistic is

$$t = \frac{(\bar{x} - \bar{y}) - \Delta}{\sqrt{\dfrac{s_1^2}{m} + \dfrac{s_2^2}{n}}} = \frac{807 - 757}{\sqrt{\dfrac{27^2}{10} + \dfrac{41^2}{10}}} = \frac{50}{\sqrt{241}} = \frac{50}{15.524} = 3.22 .$$ The approximate df is

$$v = \frac{(241)^2}{\dfrac{(72.9)^2}{9} + \dfrac{(168.1)^2}{9}} = 15.6 ,$$ which we round down to 15. The P-value for a two-tailed test is

approximately $2P(T > 3.22) = 2(.003) = .006$. This small of a P-value gives strong support for the alternative hypothesis. The data indicates a significant difference. Due to the small sample sizes (10 each), we are assuming here that compression strengths for both fixed and floating test platens are normally distributed. And, as always, we are assuming the data were randomly sampled from their respective populations.

69. Let p_1 = true proportion of returned questionnaires that included no incentive; p_2 = true proportion of returned questionnaires that included an incentive. The hypotheses are $H_0 : p_1 - p_2 = 0$ v. $H_a : p_1 - p_2 < 0$.

The test statistic is $z = \dfrac{\hat{p}_1 - \hat{p}_2}{\sqrt{\hat{p}\hat{q}\left(\frac{1}{m} + \frac{1}{n}\right)}}$.

$\hat{p}_1 = \dfrac{75}{110} = .682$ and $\hat{p}_2 = \dfrac{66}{98} = .673$; at this point, you might notice that since $\hat{p}_1 > \hat{p}_2$, the numerator of the

z statistic will be > 0 , and since we have a lower tailed test, the P-value will be $> .5$. We fail to reject H_0. This data does not suggest that including an incentive increases the likelihood of a response.

71. The center of any confidence interval for $\mu_1 - \mu_2$ is always $\bar{x}_1 - \bar{x}_2$, so $\bar{x}_1 - \bar{x}_2 = \dfrac{-473.3 + 1691.9}{2} = 609.3$.

Furthermore, half of the width of this interval is $\dfrac{1691.9 - (-473.3)}{2} = 1082.6$. Equating this value to the

expression on the right of the 95% confidence interval formula, we find $\sqrt{\dfrac{s_1^2}{n_1} + \dfrac{s_2^2}{n_2}} = \dfrac{1082.6}{1.96} = 552.35$.

For a 90% interval, the associated z value is 1.645, so the 90% confidence interval is then
$609.3 \pm (1.645)(552.35) = 609.3 \pm 908.6 = (-299.3, 1517.9)$.

73.   Let μ_1 and μ_2 denote the true mean zinc mass for Duracell and Energizer batteries, respectively. We want to test the hypotheses $H_0: \mu_1 - \mu_2 = 0$ versus $H_a: \mu_1 - \mu_2 \neq 0$. Assuming that both zinc mass distributions are normal, we'll use a two-sample t test; the test statistic is $t = \dfrac{(\bar{x} - \bar{y}) - \Delta_0}{\sqrt{\dfrac{s_1^2}{m} + \dfrac{s_2^2}{n}}} = \dfrac{(138.52 - 149.07) - 0}{\sqrt{\dfrac{(7.76)^2}{15} + \dfrac{(1.52)^2}{20}}} = -5.19.$

The textbook's formula for df gives $v = 14$. The P-value is $P(T_{14} \leq -5.19) \approx 0$. Hence, we strongly reject H_0 and we conclude the mean zinc mass content for Duracell and Energizer batteries are <u>not</u> the same (they do differ).

75.   Since we can assume that the distributions from which the samples were taken are normal, we use the two-sample t test. Let μ_1 denote the true mean headability rating for aluminum killed steel specimens and μ_2 denote the true mean headability rating for silicon killed steel. Then the hypotheses are $H_0 : \mu_1 - \mu_2 = 0$ v.

$H_a : \mu_1 - \mu_2 \neq 0$. The test statistic is $t = \dfrac{-.66}{\sqrt{.03888 + .047203}} = \dfrac{-.66}{\sqrt{.086083}} = -2.25$. The approximate

degrees of freedom are $v = \dfrac{(.086083)^2}{\dfrac{(.03888)^2}{29} + \dfrac{(.047203)^2}{29}} = 57.5 \searrow 57$. The two-tailed P-value $\approx 2(.014) = .028$,

which is less than the specified significance level, so we would reject H_0. The data supports the article's authors' claim.

77.

a.   The relevant hypotheses are $H_0 : \mu_1 - \mu_2 = 0$ v. $H_a : \mu_1 - \mu_2 \neq 0$. Assuming both populations have normal distributions, the two-sample t test is appropriate. $m = 11$, $\bar{x} = 98.1$, $s_1 = 14.2$, $n = 15$, $\bar{y} = 129.2$, $s_2 = 39.1$. The test statistic is $t = \dfrac{-31.1}{\sqrt{18.3309 + 101.9207}} = \dfrac{-31.1}{\sqrt{120.252}} = -2.84$. The

approximate degrees of freedom $v = \dfrac{(120.252)^2}{\dfrac{(18.3309)^2}{10} + \dfrac{(101.9207)^2}{14}} = 18.64 \searrow 18$. From Table A.8, the

two-tailed P-value $\approx 2(.006) = .012$. No, obviously the results are different.

b.   For the hypotheses $H_0 : \mu_1 - \mu_2 = -25$ v. $H_a : \mu_1 - \mu_2 < -25$, the test statistic changes to

$t = \dfrac{-31.1 - (-25)}{\sqrt{120.252}} = -.556$. With df $= 18$, the P-value $\approx P(T < -.6) = .278$. Since the P-value is greater than any sensible choice of α, we fail to reject H_0. There is insufficient evidence that the true average strength for males exceeds that for females by more than 25 N.

79.   To begin, we must find the % difference for each of the 10 meals! For the first meal, the % difference is $\dfrac{\text{measured} - \text{stated}}{\text{stated}} = \dfrac{212 - 180}{180} = .1778$, or 17.78%. The other nine percentage differences are 45%, 21.58%, 33.04%, 5.5%, 16.49%, 15.2%, 10.42%, 81.25%, and 26.67%.

We wish to test the hypotheses $H_0: \mu = 0$ versus $H_a: \mu \neq 0$, where μ denotes the true average percent difference for all supermarket convenience meals. A normal probability plot of these 10 values shows some noticeable deviation from linearity, so a t-test is actually of questionable validity here, but we'll proceed just to illustrate the method.

For this sample, $n = 10$, $\bar{x} = 27.29\%$, and $s = 22.12\%$, for a t statistic of $t = \dfrac{27.29 - 0}{22.12/\sqrt{10}} = 3.90$.

At df $= n - 1 = 9$, the P-value is $2P(T_9 \geq 3.90) \approx 2(.002) = .004$. Since this is smaller than any reasonable significance level, we reject H_0 and conclude that the true average percent difference between meals' stated energy values and their measured values is non-zero.

81. The normal probability plot below indicates the data for good visibility does <u>not</u> come from a normal distribution. Thus, a t-test is <u>not</u> appropriate for this small a sample size. (The plot for poor visibility isn't as bad.) That is, a pooled t test should not be used here, nor should an "unpooled" two-sample t test be used (since it relies on the same normality assumption).

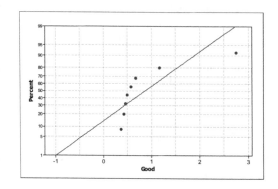

83. We wish to test H_0: $\mu_1 = \mu_2$ versus H_a: $\mu_1 \neq \mu_2$
Unpooled:

With H_0: $\mu_1 - \mu_2 = 0$ v. H_a: $\mu_1 - \mu_2 \neq 0$, we will reject H_0 if $p - value < \alpha$.

$$\nu = \dfrac{\left(\frac{.79^2}{14} + \frac{1.52^2}{12}\right)^2}{\frac{\left(\frac{.79^2}{14}\right)^2}{13} + \frac{\left(\frac{1.52^2}{12}\right)^2}{11}} = 15.95 \downarrow 15,$$ and the test statistic $t = \dfrac{8.48 - 9.36}{\sqrt{\frac{.79^2}{14} + \frac{1.52^2}{12}}} = \dfrac{-.88}{.4869} = -1.81$ leads to a P-value of about $2[P(t_{15} > 1.8)] = 2(.046) = .092$.

Pooled:

The degrees of freedom are $\nu = m + n - 2 = 14 + 12 - 2 = 24$ and the pooled variance is $\left(\dfrac{13}{24}\right)(.79)^2 + \left(\dfrac{11}{24}\right)(1.52)^2 = 1.3970$, so $s_p = 1.181$. The test statistic is $t = \dfrac{-.88}{1.181\sqrt{\frac{1}{14} + \frac{1}{12}}} = \dfrac{-.88}{.465} \approx -1.89$. The P-value $= 2[P(T_{24} > 1.9)] = 2(.035) = .070$.

With the pooled method, there are more degrees of freedom, and the P-value is smaller than with the unpooled method. That is, if we are willing to assume equal variances (which might or might not be valid here), the pooled test is more capable of detecting a significant difference between the sample means.

85.

a. With n denoting the second sample size, the first is $m = 3n$. We then wish $20 = 2(2.58)\sqrt{\dfrac{900}{3n} + \dfrac{400}{n}}$, which yields $n = 47$, $m = 141$.

b. We wish to find the n which minimizes $2z_{\alpha/2}\sqrt{\dfrac{900}{400-n} + \dfrac{400}{n}}$, or equivalently, the n which minimizes $\dfrac{900}{400-n} + \dfrac{400}{n}$. Taking the derivative with respect to n and equating to 0 yields $900(400-n)^{-2} - 400n^{-2} = 0$, whence $9n^2 = 4(400-n)^2$, or $5n^2 + 3200n - 640,000 = 0$. The solution is $n = 160$, and thus $m = 400 - n = 240$.

87. We want to test the hypothesis $H_0: \mu_1 \le 1.5\mu_2$ v. $H_a: \mu_1 > 1.5\mu_2$ – or, using the hint, $H_0: \theta \le 0$ v. $H_a: \theta > 0$.

Our point estimate of θ is $\hat{\theta} = \overline{X}_1 - 1.5\overline{X}_2$, whose estimated standard error equals $s(\hat{\theta}) = \sqrt{\dfrac{s_1^2}{n_1} + (1.5)^2\dfrac{s_2^2}{n_2}}$,

using the fact that $V(\hat{\theta}) = \dfrac{\sigma_1^2}{n_1} + (1.5)^2\dfrac{\sigma_2^2}{n_2}$. Plug in the values provided to get a test statistic $t = \dfrac{22.63 - 1.5(14.15) - 0}{\sqrt{2.8975}} \approx 0.83$. A conservative df estimate here is $v = 50 - 1 = 49$, and $t_{.05,49} \approx 1.676$. Since $0.83 < 1.676$, we fail to reject H_0 at the 5% significance level. The data does not suggest that the average tip after an introduction is more than 50% greater than the average tip without introduction.

89. $\Delta_0 = 0$, $\sigma_1 = \sigma_2 = 10$, $d = 1$, $\sigma = \sqrt{\dfrac{200}{n}} = \dfrac{14.142}{\sqrt{n}}$, so $\beta = \Phi\left(1.645 - \dfrac{\sqrt{n}}{14.142}\right)$, giving $\beta = .9015, .8264,$ .0294, and .0000 for $n = 25, 100, 2500,$ and $10,000$ respectively. If the μ_is referred to true average IQs resulting from two different conditions, $\mu_1 - \mu_2 = 1$ would have little practical significance, yet very large sample sizes would yield statistical significance in this situation.

91. $H_0: p_1 = p_2$ will be rejected at level α in favor of $H_a: p_1 > p_2$ if $z \ge z_\alpha$. With $\hat{p}_1 = \frac{250}{2500} = .10$ and $\hat{p}_2 = \frac{167}{2500} = .0668$, $\hat{p} = .0834$ and $z = \dfrac{.0332}{.0079} = 4.2$, so H_0 is rejected at any reasonable α level. It appears that a response is more likely for a white name than for a black name.

93.

a. Let μ_1 and μ_2 denote the true average weights for operations 1 and 2, respectively. The relevant hypotheses are $H_0: \mu_1 - \mu_2 = 0$ v. $H_a: \mu_1 - \mu_2 \ne 0$. The value of the test statistic is

$$t = \dfrac{(1402.24 - 1419.63)}{\sqrt{\dfrac{(10.97)^2}{30} + \dfrac{(9.96)^2}{30}}} = \dfrac{-17.39}{\sqrt{4.011363 + 3.30672}} = \dfrac{-17.39}{\sqrt{7.318083}} = -6.43 \text{ with df} =$$

$$v = \dfrac{(7.318083)^2}{\dfrac{(4.011363)^2}{29} + \dfrac{(3.30672)^2}{29}} = 57.5 \searrow 57 . \ t_{.025,57} \approx 2.000,$$ so we can reject H_0 at level .05. The data

indicates that there is a significant difference between the true mean weights of the packages for the two operations.

b. $H_0: \mu_1 = 1400$ will be tested against $H_a: \mu_1 > 1400$ using a one-sample t test with test statistic $t = \dfrac{\bar{x} - 1400}{s_1 / \sqrt{m}}$. With degrees of freedom = 29, we reject H_0 if $t \geq t_{.05,29} = 1.699$. The test statistic value is $t = \dfrac{1402.24 - 1400}{10.97 / \sqrt{30}} = \dfrac{2.24}{2.00} = 1.1$. Because $1.1 < 1.699$, H_0 is not rejected. True average weight does not appear to exceed 1400.

95. A large-sample confidence interval for $\lambda_1 - \lambda_2$ is $(\hat{\lambda}_1 - \hat{\lambda}_2) \pm z_{\alpha/2} \sqrt{\dfrac{\hat{\lambda}_1}{m} + \dfrac{\hat{\lambda}_2}{n}}$, or $(\bar{x} - \bar{y}) \pm z_{\alpha/2} \sqrt{\dfrac{\bar{x}}{m} + \dfrac{\bar{y}}{n}}$. With $\bar{x} = 1.616$ and $\bar{y} = 2.557$, the 95% confidence interval for $\lambda_1 - \lambda_2$ is $-.94 \pm 1.96(.177) = -.94 \pm .35 = (-1.29, -.59)$.

CHAPTER 10

Section 10.1

1.

    a. H_0 will be rejected if $f \geq F_{.05,4,15} = 3.06$ (since $I - 1 = 4$, and $I(J - 1) = (5)(3) = 15$). The computed value of F is $f = \dfrac{MSTr}{MSE} = \dfrac{2673.3}{1094.2} = 2.44$. Since $2.44 < 3.06$, H_0 is not rejected. The data does not indicate a difference in the mean tensile strengths of the different types of copper wires.

    b. $F_{.05,4,15} = 3.06$ and $F_{.10,4,15} = 2.36$, and our computed value of 2.44 is between those values, it can be said that $.05 < P\text{-value} < .10$.

3.   With μ_i = true average lumen output for brand i bulbs, we wish to test $H_0 : \mu_1 = \mu_2 = \mu_3$ v. H_a: at least two μ_i's are different. $MSTr = \hat{\sigma}_B^2 = \dfrac{591.2}{2} = 295.60$, $MSE = \hat{\sigma}_W^2 = \dfrac{4773.3}{21} = 227.30$, so

$$f = \frac{MSTr}{MSE} = \frac{295.60}{227.30} = 1.30.$$

For finding the P-value, we need degrees of freedom $I - 1 = 2$ and $I(J - 1) = 21$. In the 2nd row and 21st column of Table A.9, we see that $1.30 < F_{.10,2,21} = 2.57$, so the P-value $> .10$. Since .10 is not $< .05$, we cannot reject H_0. There are no statistically significant differences in the average lumen outputs among the three brands of bulbs.

5. μ_i = true mean modulus of elasticity for grade i ($i = 1, 2, 3$). We test $H_0 : \mu_1 = \mu_2 = \mu_3$ vs. H_a: at least two μ_i's are different, and we reject H_0 if $f \geq F_{.01,2,27} = 5.49$. The grand mean = 1.5367,

$$MSTr = \frac{10}{2}\left[(1.63 - 1.5367)^2 + (1.56 - 1.5367)^2 + (1.42 - 1.5367)^2\right] = .1143,$$

$$MSE = \frac{1}{3}\left[(.27)^2 + (.24)^2 + (.26)^2\right] = .0660, \quad f = \frac{MSTr}{MSE} = \frac{.1143}{.0660} = 1.73. \text{ Since } 1.73 < 5.49, \text{ we fail to reject}$$

H_0. The three grades do not appear to differ significantly.

7.   Let μ_i denote the true mean electrical resistivity for the ith mixture ($i = 1, ..., 6$).
The hypotheses are H_0: $\mu_1 = ... = \mu_6$ versus H_a: at least two of the μ_i's are different.
There are $I = 6$ different mixtures and $J = 26$ measurements for each mixture. That information provides the df values in the table. Working backwards, SSE = $I(J - 1)$MSE = 2089.350; SSTr = SST – SSE = 3575.065; MSTr = SSTr/($I - 1$) = 715.013; and, finally, f = MSTr/MSE = 51.3.

Source	df	SS	MS	f
Treatments	5	3575.065	715.013	51.3
Error	150	2089.350	13.929	
Total	155	5664.415		

The P-value is $P(F_{5,150} \geq 51.3) \approx 0$, and so H_0 will be rejected at any reasonable significance level. There is strong evidence that true mean electrical resistivity is <u>not</u> the same for all 6 mixtures.

137

9. The summary quantities are $x_{1.} = 34.3$, $x_{2.} = 39.6$, $x_{3.} = 33.0$, $x_{4.} = 41.9$, $x_{..} = 148.8$, $\Sigma\Sigma x_{ij}^2 = 946.68$, so

$$CF = \frac{(148.8)^2}{24} = 922.56, \text{ SST} = 946.68 - 922.56 = 24.12, \text{ SSTr} = \frac{(34.3)^2 + \ldots + (41.9)^2}{6} - 922.56 = 8.98,$$

$$\text{SSE} = 24.12 - 8.98 = 15.14.$$

Source	df	SS	MS	F
Treatments	3	8.98	2.99	3.95
Error	20	15.14	.757	
Total	23	24.12		

Since $3.10 = F_{.05,3,20} < 3.95 < 4.94 = F_{.01,3,20}$, $.01 < P\text{-value} < .05$, and H_0 is rejected at level .05.

Section 10.2

11. $Q_{.05,5,15} = 4.37$, $w = 4.37\sqrt{\dfrac{272.8}{4}} = 36.09$. The brands seem to divide into two groups: 1, 3, and 4; and 2 and 5; with no significant differences within each group but all between group differences are significant.

3	1	4	2	5
437.5	462.0	469.3	512.8	532.1

13. Brand 1 does not differ significantly from 3 or 4, 2 does not differ significantly from 4 or 5, 3 does not differ significantly from1, 4 does not differ significantly from 1 or 2, 5 does not differ significantly from 2, but all other differences (e.g., 1 with 2 and 5, 2 with 3, etc.) do appear to be significant.

3	1	4	2	5
427.5	462.0	469.3	502.8	532.1

15. In Exercise 10.7, $I = 6$ and $J = 26$, so the critical value is $Q_{.05,6,150} \approx Q_{.05,6,120} = 4.10$, and MSE = 13.929. So, $w \approx 4.10\sqrt{\dfrac{13.929}{26}} = 3.00$. So, sample means less than 3.00 apart will belong to the same underscored set. Three distinct groups emerge: the first mixture (in the above order), then mixtures 2-4, and finally mixtures 5-6.

14.18	17.94	18.00	18.00	25.74	27.67

17. $\theta = \Sigma c_i \mu_i$ where $c_1 = c_2 = .5$ and $c_3 = -1$, so $\hat{\theta} = .5\bar{x}_{1.} + .5\bar{x}_{2.} - \bar{x}_{3.} = -.527$ and $\Sigma c_i^2 = 1.50$. With $t_{.025,27} = 2.052$ and MSE = .0660, the desired CI is (from (10.5))

$$-.527 \pm (2.052)\sqrt{\frac{(.0660)(1.50)}{10}} = -.527 \pm .204 = (-.731, -.323).$$

19.   MSTr = 140, error df = 12, so $f = \dfrac{140}{\text{SSE}/12} = \dfrac{1680}{\text{SSE}}$ and $F_{.05,2,12} = 3.89$.

$w = Q_{.05,3,12}\sqrt{\dfrac{\text{MSE}}{J}} = 3.77\sqrt{\dfrac{\text{SSE}}{60}} = .4867\sqrt{\text{SSE}}$. Thus we wish $\dfrac{1680}{\text{SSE}} > 3.89$ (significant f) and

$.4867\sqrt{\text{SSE}} > 10$ (= 20 − 10, the difference between the extreme $\bar{x}_{i.}$'s, so no significant differences are identified). These become 431.88 > SSE and SSE > 422.16, so SSE = 425 will work.

21.

a.   Grand mean = 222.167, MSTr = 38,015.1333, MSE = 1,681.8333, and f = 22.6. The hypotheses are $H_0: \mu_1 = \dots = \mu_6$ v. H_a: at least two of the μ_i's are different. Reject H_0 if $f \geq F_{.01,5,78}$ (but since there is no table value for $v_2 = 78$, use $v_2 = 78$, use $f \geq F_{.01,5,60} = 3.34$). With $22.6 \geq 3.34$, we reject H_0. The data indicates there is a dependence on injection regimen.

b.   Assume $t_{.005,78} \approx 2.645$.

   i)   Confidence interval for $\mu_1 - \frac{1}{5}(\mu_2 + \mu_3 + \mu_4 + \mu_5 + \mu_6)$: $\Sigma c_i \bar{x}_i \pm t_{\alpha/2,I(J-1)}\sqrt{\dfrac{\text{MSE}\left(\Sigma c_i^2\right)}{J}}$

$= -67.4 \pm (2.645)\sqrt{\dfrac{1,681.8333(1.2)}{14}} = (-99.16, -35.64)$.

   ii)   Confidence interval for $\frac{1}{4}(\mu_2 + \mu_3 + \mu_4 + \mu_5) - \mu_6$:

$= 61.75 \pm (2.645)\sqrt{\dfrac{1,681.8333(1.25)}{14}} = (29.34, 94.16)$

Section 10.3

23. $J_1 = 5, J_2 = 4, J_3 = 4, J_4 = 5, \bar{x}_{1.} = 58.28, \bar{x}_{2.} = 55.40, \bar{x}_{3.} = 50.85, \bar{x}_{4.} = 45.50$, MSE = 8.89.

With $W_{ij} = Q_{.05,4,14} \cdot \sqrt{\dfrac{\text{MSE}}{2}\left(\dfrac{1}{J_i} + \dfrac{1}{J_j}\right)} = 4.11\sqrt{\dfrac{8.89}{2}\left(\dfrac{1}{J_i} + \dfrac{1}{J_j}\right)}$,

$\bar{x}_{1.} - \bar{x}_{2.} \pm W_{12} = (2.88) \pm (5.81)$; $\bar{x}_{1.} - \bar{x}_{3.} \pm W_{13} = (7.43) \pm (5.81)$ *; $\bar{x}_{1.} - \bar{x}_{4.} \pm W_{14} = (12.78) \pm (5.48)$ *;

$\bar{x}_{2.} - \bar{x}_{3.} \pm W_{23} = (4.55) \pm (6.13)$; $\bar{x}_{2.} - \bar{x}_{4.} \pm W_{24} = (9.90) \pm (5.81)$ *; $\bar{x}_{3.} - \bar{x}_{4.} \pm W_{34} = (5.35) \pm (5.81)$.

A * indicates an interval that doesn't include zero, corresponding to μ's that are judged significantly different. This underscoring pattern does not have a very straightforward interpretation.

4   3   2   1

25.

a. The distributions of the polyunsaturated fat percentages for each of the four regimens must be normal with equal variances.

b. We have all the $\bar{x}_i$s, and we need the grand mean:

$$\bar{x}_{..} = \frac{8(43.0)+13(42.4)+17(43.1)+14(43.5)}{52} = \frac{2236.9}{52} = 43.017;$$

$$\text{SSTr} = \sum J_i(\bar{x}_{i.} - \bar{x}_{..})^2 = 8(43.0-43.017)^2 + 13(42.4-43.017)^2$$

$$+ 17(43.1-43.017)^2 + 13(43.5-43.017)^2 = 8.334 \text{ and MSTr} = \frac{8.334}{3} = 2.778$$

$$\text{SSE} = \sum(J_i-1)s^2 = 7(1.5)^2 + 12(1.3)^2 + 16(1.2)^2 + 13(1.2)^2 = 77.79 \text{ and } \text{MSE} = \frac{77.79}{48} = 1.621. \text{ Then}$$

$$f = \frac{\text{MSTr}}{\text{MSE}} = \frac{2.778}{1.621} = 1.714$$

Since $1.714 < F_{.10,3,50} = 2.20$, we can say that the P-value is $> .10$. We do not reject the null hypothesis at significance level .10 (or any smaller), so we conclude that the data suggests no difference in the percentages for the different regimens.

27.

a. Let μ_i = true average folacin content for specimens of brand i. The hypotheses to be tested are $H_0 : \mu_1 = \mu_2 = \mu_3 = \mu_4$ vs. H_a: at least two of the μ_i's are different . $\Sigma\Sigma x_{ij}^2 = 1246.88$ and

$$\frac{x_{..}^2}{n} = \frac{(168.4)^2}{24} = 1181.61, \text{ so SST} = 65.27; \quad \frac{\Sigma x_{i.}^2}{J_i} = \frac{(57.9)^2}{7} + \frac{(37.5)^2}{5} + \frac{(38.1)^2}{6} + \frac{(34.9)^2}{6} = 1205.10, \text{ so}$$

$\text{SSTr} = 1205.10 - 1181.61 = 23.49$.

Source	df	SS	MS	F
Treatments	3	23.49	7.83	3.75
Error	20	41.78	2.09	
Total	23	65.27		

With numerator df = 3 and denominator df = 20, $F_{.05,3,20} = 3.10 < 3.75 < F_{.01,3,20} = 4.94$, so the P-value is between .01 and .05. We reject H_0 at the .05 level: at least one of the pairs of brands of green tea has different average folacin content.

b. With $\bar{x}_{i.}$ = 8.27, 7.50, 6.35, and 5.82 for i = 1, 2, 3, 4, we calculate the residuals $x_{ij} - \bar{x}_{i.}$ for all observations. A normal probability plot appears below and indicates that the distribution of residuals could be normal, so the normality assumption is plausible. The sample standard deviations are 1.463, 1.681, 1.060, and 1.551, so the equal variance assumption is plausible (since the largest sd is less than twice the smallest sd).

Normal Probability Plot for ANOVA Residuals

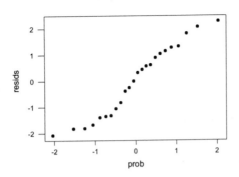

c. $Q_{.05,4,20} = 3.96$ and $W_{ij} = 3.96 \cdot \sqrt{\dfrac{2.09}{2}\left(\dfrac{1}{J_i}+\dfrac{1}{J_j}\right)}$, so the Modified Tukey intervals are:

Pair	Interval	Pair	Interval
1,2	$.77 \pm 2.37$	2,3	$1.15 \pm 2.45a$
1,3	1.92 ± 2.25	2,4	1.68 ± 2.45
1,4	2.45 ± 2.25 *	3,4	$.53 \pm 2.34$

$$\underline{\quad 4 \qquad 3 \qquad 2 \qquad 1 \quad}$$

Only Brands 1 and 4 are significantly different from each other.

29. $E(\text{SSTr}) = E\left(\sum_i J_i \bar{X}_{i\cdot}^2 - n\bar{X}_{\cdot\cdot}^2\right) = \sum J_i E\left(\bar{X}_{i\cdot}^2\right) - nE\left(\bar{X}_{\cdot\cdot}^2\right)$

$= \sum J_i \left[V\left(\bar{X}_{i\cdot}\right)+\left(E\left(\bar{X}_{i\cdot}\right)\right)^2\right] - n\left[V\left(\bar{X}_{\cdot\cdot}\right)+\left(E\left(\bar{X}_{\cdot\cdot}\right)\right)^2\right] = \sum J_i\left[\dfrac{\sigma^2}{J_i}+\mu_i^2\right] - n\left[\dfrac{\sigma^2}{n}+\left(\dfrac{\sum J_i \mu_i}{n}\right)^2\right]$

$= I\sigma^2 + \sum J_i \left(\mu+\alpha_i\right)^2 - \sigma^2 - \dfrac{1}{n}\left[\sum J_i \left(\mu+\alpha_i\right)\right]^2 = (I-1)\sigma^2 + \sum J_i \mu^2 + 2\mu\sum J_i \alpha_i + \sum J_i \alpha_i^2 - \dfrac{1}{n}\left[n\mu+0\right]^2$

$= (I-1)\sigma^2 + \mu^2 n + 2\mu 0 + \sum J_i \alpha_i^2 - n\mu^2 = (I-1)\sigma^2 + \sum J_i \alpha_i^2$, from which $E(\text{MSTr})$ is obtained through division by $(I-1)$.

31. With $\sigma = 1$ (any other σ would yield the same ϕ), $\alpha_1 = -1$, $\alpha_2 = \alpha_3 = 0$, $\alpha_4 = 1$,

$\phi^2 = \dfrac{1\left(5(-1)^2 + 4(0)^2 + 4(0)^2 + 5(1)^2\right)}{4} = 2.5$, $\phi = 1.58$, $v_1 = 3$, $v_2 = 14$, and power $\approx .65$.

33. $g(x) = x\left(1-\dfrac{x}{n}\right) = nu(1-u)$ where $u = \dfrac{x}{n}$, so $h(x) = \int \left[u(1-u)\right]^{-1/2} du$. From a table of integrals, this gives $h(x) = \arcsin\left(\sqrt{u}\right) = \arcsin\left(\sqrt{\dfrac{x}{n}}\right)$ as the appropriate transformation.

141

Supplementary Exercises

35.

    a. $H_0 : \mu_1 = \mu_2 = \mu_3 = \mu_4$ v. H_a: at least two of the μ_i's are different; $f = 3.68 < F_{.01,3,20} = 4.94$, thus fail to reject H_0. The means do not appear to differ.

    b. We reject H_0 when the P-value $< \alpha$. Since .029 is not $< .01$, we still fail to reject H_0.

37. Let μ_i = true average amount of motor vibration for each of five bearing brands. Then the hypotheses are $H_0 : \mu_1 = ... = \mu_5$ vs. H_a: at least two of the μ_i's are different. The ANOVA table follows:

Source	df	SS	MS	F
Treatments	4	30.855	7.714	8.44
Error	25	22.838	0.914	
Total	29	53.694		

$8.44 > F_{.001,4,25} = 6.49$, so P-value $< .001 < .05$, so we reject H_0. At least two of the means differ from one another. The Tukey multiple comparisons are appropriate. $Q_{.05,5,25} = 4.15$ from Minitab output; or, using Table A.10, we can approximate with $Q_{.05,5,24} = 4.17$. $W_{ij} = 4.15\sqrt{.914/6} = 1.620$.

Pair	$\overline{x}_{i.} - \overline{x}_{j.}$	Pair	$\overline{x}_{i.} - \overline{x}_{j.}$
1,2	-2.267*	2,4	1.217
1,3	0.016	2,5	2.867*
1,4	-1.050	3,4	-1.066
1,5	0.600	3,5	0.584
2,3	2.283*	4,5	1.650*

    *Indicates significant pairs.

$$\underline{\quad 5 \qquad 3 \qquad 1 \qquad 4 \qquad 2 \quad}$$

39. $\hat{\theta} = 2.58 - \dfrac{2.63 + 2.13 + 2.41 + 2.49}{4} = .165$, $t_{.025,25} = 2.060$, MSE = .108, and

$\Sigma c_i^2 = (1)^2 + (-.25)^2 + (-.25)^2 + (-.25)^2 + (-.25)^2 = 1.25$, so a 95% confidence interval for θ is

$.165 \pm 2.060\sqrt{\dfrac{(.108)(1.25)}{6}} = .165 \pm .309 = (-.144,.474)$. This interval does include zero, so 0 is a plausible value for θ.

41. This is a random effects situation. $H_0 : \sigma_A^2 = 0$ states that variation in laboratories doesn't contribute to variation in percentage. H_0 will be rejected in favor of H_a if $f \geq F_{.05,3,8} = 4.07$. SST = 86,078.9897 − 86,077.2224 = 1.7673, SSTr = 1.0559, and SSE = .7114. Thus $f = 3.96 < 4.07$, so H_0 cannot be rejected at level .05. Variation in laboratories does not appear to be present.

43. $\sqrt{(I-1)(\text{MSE})\left(F_{.05, I-1, n-I}\right)} = \sqrt{(2)(2.39)(3.63)} = 4.166$. For $\mu_1 - \mu_2$, $c_1 = 1$, $c_2 = -1$, and $c_3 = 0$, so

$\sqrt{\sum \dfrac{c_i^2}{J_i}} = \sqrt{\dfrac{1}{8} + \dfrac{1}{5}} = .570$. Similarly, for $\mu_1 - \mu_3$, $\sqrt{\sum \dfrac{c_i^2}{J_i}} = \sqrt{\dfrac{1}{8} + \dfrac{1}{6}} = .540$; for $\mu_2 - \mu_3$,

$\sqrt{\sum \dfrac{c_i^2}{J_i}} = \sqrt{\dfrac{1}{5} + \dfrac{1}{6}} = .606$, and for $.5\mu_2 + .5\mu_2 - \mu_3$, $\sqrt{\sum \dfrac{c_i^2}{J_i}} = \sqrt{\dfrac{.5^2}{8} + \dfrac{.5^2}{5} + \dfrac{(-1)^2}{6}} = .498$.

Contrast	Estimate	Interval
$\mu_1 - \mu_2$	$25.59 - 26.92 = -1.33$	$(-1.33) \pm (.570)(4.166) = (-3.70, 1.04)$
$\mu_1 - \mu_3$	$25.59 - 28.17 = -2.58$	$(-2.58) \pm (.540)(4.166) = (-4.83, -.33)$
$\mu_2 - \mu_3$	$26.92 - 28.17 = -1.25$	$(-1.25) \pm (.606)(4.166) = (-3.77, 1.27)$
$.5\mu_2 + .5\mu_2 - \mu_3$	-1.92	$(-1.92) \pm (.498)(4.166) = (-3.99, 0.15)$

The contrast between μ_1 and μ_3, since the calculated interval is the only one that does not contain 0.

45. $Y_{ij} - \overline{Y}_{..} = c\left(X_{ij} - \overline{X}_{..}\right)$ and $\overline{Y}_{i.} - \overline{Y}_{..} = c\left(\overline{X}_{i.} - \overline{X}_{..}\right)$, so each sum of squares involving Y will be the corresponding sum of squares involving X multiplied by c^2. Since F is a ratio of two sums of squares, c^2 appears in both the numerator and denominator. So c^2 cancels, and F computed from Y_{ij}'s = F computed from X_{ij}'s.

CHAPTER 11

Section 11.1

1.

a. $MSA = \dfrac{30.6}{4} = 7.65$, $MSE = \dfrac{59.2}{12} = 4.93$, $f_A = \dfrac{7.65}{4.93} = 1.55$. Since 1.55 is not $\geq F_{.05,4,12} = 3.26$, don't reject H_{0A}. There is no significant difference in true average tire lifetime due to different makes of cars.

b. $MSB = \dfrac{44.1}{3} = 14.70$, $f_B = \dfrac{14.70}{4.93} = 2.98$. Since 2.98 is not $\geq F_{.05,3,12} = 3.49$, don't reject H_{0B}. There is no significant difference in true average tire lifetime due to different brands of tires.

3. $x_{1\bullet} = 927$, $x_{2\bullet} = 1301$, $x_{3\bullet} = 1764$, $x_{4\bullet} = 2453$, $x_{\bullet 1} = 1347$, $x_{\bullet 2} = 1529$, $x_{\bullet 3} = 1677$,

$x_{\bullet 4} = 1892$, $x_{\bullet\bullet} = 6445$, $\Sigma\Sigma x_{ij}^2 = 2{,}969{,}375$, $CF = \dfrac{(6445)^2}{16} = 2{,}596{,}126.56$,

$SSA = 324{,}082.2$, $SSB = 39{,}934.2$, $SST = 373{,}248.4$, $SSE = 9232.0$

a.

Source	Df	SS	MS	F
A	3	324,082.2	108,027.4	105.3
B	3	39,934.2	13,311.4	13.0
Error	9	9232.0	1025.8	
Total	15	373,248.4		

Since $F_{.01,3,9} = 6.99$, both H_{0A} and H_{0B} are rejected.

b. $Q_{.01,4,9} = 5.96$, $w = 5.96\sqrt{\dfrac{1025.8}{4}} = 95.4$

i:	1	2	3	4
$\bar{x}_{i\bullet}$:	231.75	325.25	441.00	613.25

All levels of Factor A (gas rate) differ significantly except for 1 and 2

c. $w = 95.4$, as in **b**

i:	1	2	3	4
$\bar{x}_{\bullet j}$:	336.75	382.25	419.25	473

144

Only levels 1 and 4 appear to differ significantly.

5.

Source	Df	SS	MS	F
Angle	3	58.16	19.3867	2.5565
Connector	4	246.97	61.7425	8.1419
Error	12	91.00	7.5833	
Total	19	396.13		

$H_0 : \alpha_1 = \alpha_2 = \alpha_3 = \alpha_4 = 0$; H_a : at least one α_i is not zero.

$f_A = 2.5565 < F_{.01,3,12} = 5.95$, so fail to reject H_0. The data fails to indicate any effect due to the angle of pull.

7.

a. SSE = SST − SSTr − SSBl = 3476.00 − 28.78 − 2977.67 = 469.55; MSTr = 28.78/(3−1) = 14.39, MSE = SSE/[(18−1)(3−1)] = 13.81. Hence, f_{Tr} = 1.04, which is clearly insignificant when compared to $F_{.05,2,34}$.

b. $f_{Bl} = 12.68$, which is significant, and suggests substantial variation among subjects. If we had not controlled for such variation, it might have affected the analysis and conclusions.

9.

Source	Df	SS	MS	f
Treatment	3	81.1944	27.0648	22.36
Block	8	66.5000	8.3125	6.87
Error	24	29.0556	1.2106	
Total	35	176.7500		

$F_{.05,3,24} = 3.01$. Reject H_0. There is an effect due to treatments.

$Q_{.05,4,24} = 3.90$; $w = (3.90)\sqrt{\dfrac{1.2106}{9}} = 1.43$

1	4	3	2
8.56	9.22	10.78	12.44

11. The residual, percentile pairs are $(-0.1225, -1.73)$, $(-0.0992, -1.15)$, $(-0.0825, -0.81)$, $(-0.0758, -0.55)$, $(-0.0750, -0.32)$, $(0.0117, -0.10)$, $(0.0283, 0.10)$, $(0.0350, 0.32)$, $(0.0642, 0.55)$, $(0.0708, 0.81)$, $(0.0875, 1.15)$, $(0.1575, 1.73)$.

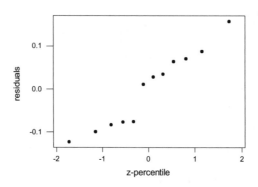

The pattern is sufficiently linear, so normality is plausible.

13.

a. With $Y_{ij} = X_{ij} + d$, $\overline{Y}_{i\bullet} = \overline{X}_{i\bullet} + d$, $\overline{Y}_{\bullet j} = \overline{X}_{\bullet j} + d$, $\overline{Y}_{\bullet\bullet} = \overline{X}_{\bullet\bullet} + d$, so all quantities inside the parentheses in (11.5) remain unchanged when the Y quantities are substituted for the corresponding X's (e.g., $\overline{Y}_{i\bullet} - \overline{Y}_{\bullet\bullet} = \overline{X}_{i\bullet} - \overline{X}_{\bullet\bullet}$, etc.).

b. With $Y_{ij} = cX_{ij}$, each sum of squares for Y is the corresponding SS for X multiplied by c^2. However, when F ratios are formed the c^2 factors cancel, so all F ratios computed from Y are identical to those computed from X. If $Y_{ij} = cX_{ij} + d$, the conclusions reached from using the Y's will be identical to those reached using the X's.

15.

a. $\Sigma\alpha_i^2 = 24$, so $\phi^2 = \left(\dfrac{3}{4}\right)\left(\dfrac{24}{16}\right) = 1.125$, $\phi = 1.06$, $\nu_1 = 3$, $\nu_2 = 6$, and from figure 10.5, power $\approx .2$. For the second alternative, $\phi = 1.59$, and power $\approx .43$.

b. $\phi^2 = \left(\dfrac{I}{J}\right)\Sigma\dfrac{\beta_j^2}{\sigma^2} = \left(\dfrac{4}{5}\right)\left(\dfrac{20}{16}\right) = 1.00$, so $\phi = 1.00$, $\nu_1 = 4$, $\nu_2 = 12$, and power $\approx .3$.

Section 11.2

17.

a.

Source	Df	SS	MS	f	$F_{.05}$
Sand	2	705	352.5	3.76	4.26
Fiber	2	1,278	639.0	6.82*	4.26
Sand&Fiber	4	279	69.75	0.74	3.63
Error	9	843	93.67		
Total	17	3,105			

There appears to be an effect due to carbon fiber addition.

b.

Source	Df	SS	MS	f	$F_{.05}$
Sand	2	106.78	53.39	6.54*	4.26
Fiber	2	87.11	43.56	5.33*	4.26
Sand&Fiber	4	8.89	2.22	.27	3.63
Error	9	73.50	8.17		
Total	17	276.28			

There appears to be an effect due to both sand and carbon fiber addition to casting hardness.

c.

Sand%	0	15	30	0	15	30	0	15	30
Fiber%	0	0	0	0.25	0.25	0.25	0.5	0.5	0.5
$\bar{x}$	62	68	69.5	69	71.5	73	68	71.5	74

The plot below indicates some effect due to sand and fiber addition with no significant interaction. This agrees with the statistical analysis in part **b**.

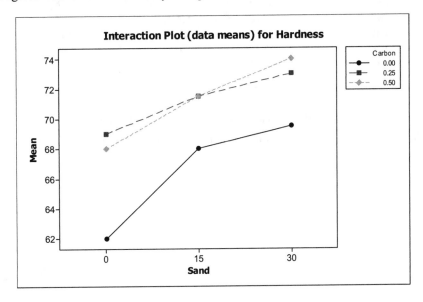

147

19.

a. Software gives the ANOVA table below. Call coal type Factor A and NaOH concentration Factor B. H_{0AB} cannot be rejected ($f = 0.21$, P-value = .924), so no significant interaction is present. H_{0B} cannot be rejected ($f = 3.66$, P-value = .069), so varying levels of NaOH does not have a significant impact on total acidity. H_{0A} is rejected at $\alpha = .01$ ($f = 29.49$, P-value = .000): type of coal does appear to affect total acidity.

Two-way ANOVA: Acidity versus Coal, NaOH

```
Source        DF      SS        MS       F       P
Coal          2   1.00241  0.501206  29.49   0.000
NaOH          2   0.12431  0.062156   3.66   0.069
Interaction   4   0.01456  0.003639   0.21   0.924
Error         9   0.15295  0.016994
Total        17   1.29423
```

b. $Q_{.01,3,9} = 5.43$, $w = 5.43\sqrt{\dfrac{.0170}{6}} = .289$

i:	3	1	2
$\overline{x}_{i\cdot\cdot}$	8.035	8.247	8.607

Coal 2 is judged significantly different from both 1 and 3, but these latter two don't differ significantly from each other.

21. From the provided SS, $SSAB = 64,954.70 - [22,941.80 + 22,765.53 + 15,253.50] = 3993.87$. This allows us to complete the ANOVA table below.

Source	df	SS	MS	f
A	2	22,941.80	11,470.90	22.98
B	4	22,765.53	5691.38	11.40
AB	8	3993.87	499.23	.49
Error	15	15,253.50	1016.90	
Total	29	64,954.70		

$f_{AB} = .49$ is clearly not significant. Since $22.98 \geq F_{.05,2,8} = 4.46$, H_{0A} is rejected; since $11.40 \geq F_{.05,4,8} = 3.84$, H_{0B} is also rejected. We conclude that the different cement factors affect flexural strength differently and that batch variability contributes to variation in flexural strength.

23. Summary quantities include $x_{1\bullet\bullet} = 9410$, $x_{2\bullet\bullet} = 8835$, $x_{3\bullet\bullet} = 9234$, $x_{\bullet 1\bullet} = 5432$, $x_{\bullet 2\bullet} = 5684$, $x_{\bullet 3\bullet} = 5619$, $x_{\bullet 4\bullet} = 5567$, $x_{\bullet 3\bullet} = 5177$, $x_{\bullet\bullet\bullet} = 27{,}479$, $CF = 16{,}779{,}898.69$, $\Sigma x_{i\bullet\bullet}^2 = 251{,}872{,}081$, $\Sigma x_{\bullet j\bullet}^2 = 151{,}180{,}459$, resulting in the accompanying ANOVA table.

Source	df	SS	MS	f
A	2	11,573.38	5786.69	$\frac{MSA}{MSAB} = 26.70$
B	4	17,930.09	4482.52	$\frac{MSB}{MSAB} = 20.68$
AB	8	1734.17	216.77	$\frac{MSAB}{MSE} = 1.38$
Error	30	4716.67	157.22	
Total	44	35,954.31		

Since $1.38 < F_{.01,8,30} = 3.17$, H_{0G} cannot be rejected, and we continue: $26.70 \geq F_{.01,2,8} = 8.65$, and $20.68 \geq F_{.01,4,8} = 7.01$, so both H_{0A} and H_{0B} are rejected. Both capping material and the different batches affect compressive strength of concrete cylinders.

25. With $\theta = \alpha_i - \alpha_i'$, $\hat{\theta} = \overline{X}_{i..} - \overline{X}_{i'..} = \frac{1}{JK}\Sigma\Sigma\left(X_{ijk} - X_{i'jk}\right)$, and since $i \neq i'$, X_{ijk} and $X_{i'jk}$ are independent for every j, k. Thus, $V(\hat{\theta}) = V(\overline{X}_{i..}) + V(\overline{X}_{i'..}) = \frac{\sigma^2}{JK} + \frac{\sigma^2}{JK} = \frac{2\sigma^2}{JK}$ (because $V(\overline{X}_{i..}) = V(\overline{\varepsilon}_{i..})$ and $V(\varepsilon_{ijk}) = \sigma^2$) so $\hat{\sigma}_{\hat{\theta}} = \sqrt{\frac{2MSE}{JK}}$. The appropriate number of df is $IJ(K-1)$, so the CI is $\left(\overline{x}_{i..} - \overline{x}_{i'..}\right) \pm t_{\alpha/2,IJ(K-1)}\sqrt{\frac{2MSE}{JK}}$. For the data of exercise 19, $\overline{x}_{2..} = 8.192$, $\overline{x}_{3..} = 8.395$, $MSE = .0170$, $t_{.025,9} = 2.262$, $J = 3$, $K = 2$, so the 95% C.I. for $\alpha_2 - \alpha_3$ is $(8.182 - 8.395) \pm 2.262\sqrt{\frac{.0340}{6}} = -0.203 \pm 0.170 = (-0.373, -0.033)$.

Section 11.3

27.

a.

Source	Df	SS	MS	f	$F_{.05}$
A	2	14,144.44	7072.22	61.06	3.35
B	2	5,511.27	2755.64	23.79	3.35
C	2	244,696.39	122.348.20	1056.24	3.35
AB	4	1,069.62	267.41	2.31	2.73
AC	4	62.67	15.67	.14	2.73
BC	4	331.67	82.92	.72	2.73
ABC	8	1,080.77	135.10	1.17	2.31
Error	27	3,127.50	115.83		
Total	53	270,024.33			

b. The computed F-statistics for all four interaction terms are less than the tabled values for statistical significance at the level .05. This indicates that none of the interactions are statistically significant.

c. The computed F-statistics for all three main effects exceed the tabled value for significance at level .05. All three main effects are statistically significant.

d. Since $Q_{.05,3,27}$ is not tabled, use $Q_{.05,3,24} = 3.53$, $w = 3.53\sqrt{\dfrac{115.83}{(3)(3)(2)}} = 8.95$. All three levels differ

significantly from each other.

29. $I = 3, J = 2, K = 4, L = 4$; $\bar{x}_{1...} = 3.781$; $SSB = IKL\sum(\bar{x}_{.j..} - \bar{x}_{....})^2$; $SSC = IJL\sum(\bar{x}_{..k.} - \bar{x}_{....})^2$.

For level A: $\bar{x}_{1...} = 3.781$ $\bar{x}_{2...} = 3.625$ $\bar{x}_{3...} = 4.469$

For level B: $\bar{x}_{.1..} = 4.979$ $\bar{x}_{.2..} = 2.938$

For level C: $\bar{x}_{..1.} = 3.417$ $\bar{x}_{..2.} = 5.875$ $\bar{x}_{..3.} = .875$ $\bar{x}_{..4.} = 5.667$

         $\bar{x}_{....} = 3.958$

SSA = 12.907;   SSB = 99.976;   SSC = 393.436

a.

Source	Df	SS	MS	f	$F_{.05}$*
A	2	12.907	6.454	1.04	3.15
B	1	99.976	99.976	16.09	4.00
C	3	393.436	131.145	21.10	2.76
AB	2	1.646	.823	.13	3.15
AC	6	71.021	11.837	1.90	2.25
BC	3	1.542	.514	.08	2.76
ABC	6	9.805	1.634	.26	2.25
Error	72	447.500	6.215		
Total	95	1,037.833			

*use 60 df for denominator of tabled F.

b. No interaction effects are significant at level .05.

c. Factor B and C main effects are significant at the level .05.

d. $Q_{.05,4,72}$ is not tabled, use $Q_{.05,4,60} = 3.74$, $w = 3.74\sqrt{\dfrac{6.215}{(3)(2)(4)}} = 1.90$.

       Machine: 3   1   4   2

         Mean: .875   3.417   5.667   5.875

31.

$x_{ij.}$	B_1	B_2	B_3
A_1	210.2	224.9	218.1
A_2	224.1	229.5	221.5
A_3	217.7	230.0	202.0
$x_{.j.}$	652.0	684.4	641.6

$x_{i.k}$	A_1	A_2	A_3
C_1	213.8	222.0	205.0
C_2	225.6	226.5	223.5
C_3	213.8	226.6	221.2
$x_{i..}$	653.2	675.1	649.7

$x_{.jk}$	C_1	C_2	C_3
B_1	213.5	220.5	218.0
B_2	214.3	246.1	224.0
B_3	213.0	209.0	219.6
$x_{..k}$	640.8	675.6	661.6

$\Sigma\Sigma x_{ij.}^2 = 435,382.26$ $\Sigma\Sigma x_{i.k}^2 = 435,156.74$ $\Sigma\Sigma x_{.jk}^2 = 435,666.36$

$\Sigma x_{.j.}^2 = 1,305,157.92$ $\Sigma x_{i..}^2 = 1,304,540.34$ $\Sigma x_{..k}^2 = 1,304,774.56$

Also, $\Sigma\Sigma\Sigma x_{ijk}^2 = 145,386.40$, $x_{...} = 1978$, CF = 144,906.81, from which we obtain the ANOVA table displayed in the problem statement. $F_{.01,4,8} = 7.01$, so the AB and BC interactions are significant (as can be seen from the P-values) and tests for main effects are not appropriate.

33.

Source	Df	SS	MS	f
A	6	67.32	11.02	
B	6	51.06	8.51	
C	6	5.43	.91	.61
Error	30	44.26	1.48	
Total	48	168.07		

Since $.61 < F_{.05,6,30} = 2.42$, treatment was not effective.

35.

	1	2	3	4	5	
$x_{i..}$	40.68	30.04	44.02	32.14	33.21	$\Sigma x_{i..}^2 = 6630.91$
$x_{.j.}$	29.19	31.61	37.31	40.16	41.82	$\Sigma x_{.j.}^2 = 6605.02$
$x_{..k}$	36.59	36.67	36.03	34.50	36.30	$\Sigma x_{..k}^2 = 6489.92$

$x_{...} = 180.09$, CF = 1297.30, $\Sigma\Sigma x_{ij(k)}^2 = 1358.60$

Source	df	SS	MS	f
A	4	28.89	7.22	10.71
B	4	23.71	5.93	8.79
C	4	0.63	0.16	0.23
Error	12	8.09	0.67	
Total	24	61.30		

$F_{.05,4,12} = 3.26$, so both factor A (plant) and B(leaf size) appear to affect moisture content, but factor C (time of weighing) does not.

37.

Source	Df	MS	f	$F_{.01}$*
A	2	2207.329	2259.29	5.39
B	1	47.255	48.37	7.56
C	2	491.783	503.36	5.39
D	1	.044	.05	7.56
AB	2	15.303	15.66	5.39
AC	4	275.446	281.93	4.02
AD	2	.470	.48	5.39
BC	2	2.141	2.19	5.39
BD	1	.273	.28	7.56
CD	2	.247	.25	5.39
ABC	4	3.714	3.80	4.02
ABD	2	4.072	4.17	5.39
ACD	4	.767	.79	4.02
BCD	2	.280	.29	5.39
ABCD	4	.347	.355	4.02
Error	36	.977		
Total	71			

*Because denominator df for 36 is not tabled, use df = 30.

SST = (71)(93.621) = 6,647.091. Computing all other sums of squares and adding them up = 6,645.702.
Thus SSABCD = 6,647.091 – 6,645.702 = 1.389 and $MSABCD = \dfrac{1.389}{4} = .347$.

At level .01 the statistically significant main effects are A, B, C. The interaction AB and AC are also statistically significant. No other interactions are statistically significant.

Section 11.4

39.

Condition	Total	1	2	Contrast	$SS = \frac{(contrast)^2}{24}$
111	315	927	2478	5485	
211	612	1551	3007	1307	A = 71,177.04
121	584	1163	680	1305	B = 70,959.38
221	967	1844	627	199	AB = 1650.04
112	453	297	624	529	C = 11,660.04
212	710	383	681	−53	AC = 117.04
122	737	257	86	57	BC = 135.38
222	1107	370	113	27	ABC = 30.38

a. $\hat{\beta}_1 = \bar{x}_{.2..} - \bar{x}_{....} = \dfrac{584 + 967 + 737 + 1107 - 315 - 612 - 453 - 710}{24} = 54.38$

$\hat{\gamma}_{11}^{AC} = \dfrac{315 - 612 + 584 - 967 - 453 + 710 - 737 + 1107}{24} = 2.21; \ \hat{\gamma}_{21}^{AC} = -\hat{\gamma}_{11}^{AC} = 2.21.$

b. Factor SS's appear above. With $CF = \dfrac{5485^2}{24} = 1{,}253{,}551.04$ and $\Sigma\Sigma\Sigma\Sigma x_{ijkl}^2 = 1{,}411{,}889$, SST

$= 158{,}337.96$, from which SSE $= 2608.7$. The ANOVA table appears in the answer section.

$F_{.05,1,16} = 4.49$, from which we see that the AB interaction and al the main effects are significant.

41. $\Sigma\Sigma\Sigma\Sigma\Sigma x_{ijklm}^2 = 3{,}308{,}143$, $x_{.....} = 11{,}956$, so $CF = \dfrac{(11{,}956)^2}{48} = 2{,}979{,}535.02$, and SST $= 328{,}607.98$. Each

SS is $\dfrac{(\text{effect contrast})^2}{48}$ and SSE is obtained by subtraction. The ANOVA table appears in the answer

section (see back of textbook). $F_{.05,1,32} \approx 4.15$, a value exceeded by the F ratios for AB interaction and the

four main effects.

154

43.

Condition/ Effect	$SS = \frac{(contrast)^2}{16}$	f	Condition/ Effect	$SS = \frac{(contrast)^2}{16}$	f
(1)	—		D	414.123	850.77
A	.436	< 1	AD	.017	< 1
B	.099	< 1	BD	.456	< 1
AB	.003	< 1	ABD	.990	—
C	.109	< 1	CD	2.190	4.50
AC	.078	< 1	ACD	1.020	—
BC	1.404	3.62	BCD	.133	—
ABC	.286	—	ABCD	.004	—

SSE = .286 + .990 + 1.020 + .133 + .004 =2.433, df = 5, so MSE = .487, which forms the denominators of the f values above. $F_{.05,1,5} = 6.61$, so only the D main effect is significant.

45.

a. The allocation of treatments to blocks is as given in the answer section (see back of book), with block #1 containing all treatments having an even number of letters in common with both *ab* and *cd*, block #2 those having an odd number in common with *ab* and an even number with *cd*, etc.

b. $\Sigma\Sigma\Sigma\Sigma x_{ijklm}^2 = 9,035,054$ and $x_{.....} = 16,898$, so $SST = 9,035,054 - \frac{16,898^2}{32} = 111,853.875$. The eight

block-replication totals are 2091 (= 618 + 421 + 603 + 449, the sum of the four observations in block #1 on replication #1), 2092, 2133, 2145, 2113, 2080, 2122, and 2122, so

$SSBl = \frac{2091^2}{4} + ... + \frac{2122^2}{4} - \frac{16,898^2}{32} = 898.875$. The effect SS's can be computed via Yates'

algorithm; those we keep appear below. SSE is computed by SST – [sum of all other SS]. MSE = 5475.75/12 = 456.3125, which forms the denominator of the F ratios below. With $F_{.01,1,12} = 9.33$, only

the A and B main effects are significant.

Source	df	SS	f
A	1	12403.125	27.18
B	1	92235.125	202.13
C	1	3.125	0.01
D	1	60.500	0.13
AC	1	10.125	0.02
BC	1	91.125	0.20
AD	1	50.000	0.11
BC	1	420.500	0.92
ABC	1	3.125	0.01
ABD	1	0.500	0.00
ACD	1	200.000	0.44
BCD	1	2.000	0.00
Block	7	898.875	0.28
Error	12	5475.750	
Total	31	111853.875	

155

47.

a. The third nonestimable effect is $(ABCDE)(CDEFG) = ABFG$. The treatments in the group containing (1) are (1), *ab, cd, ce, de, fg, acf, adf, adg, aef, acg, aeg, bcg, bcf, bdf, bdg, bef, beg, abcd, abce, abde, abfg, cdfg, cefg, defg, acdef, acdeg, bcdef, bcdeg, abcdfg, abcefg, abdefg*. The alias groups of the seven main effects are {A, $BCDE$, $ACDEFG$, BFG}, {B, $ACDE$, $BCDEFG$, AFG}, {C, $ABDE$, $DEFG$, $ABCFG$}, {D, $ABCE$, $CEFG$, $ABDFG$}, {E, $ABCD$, $CDFG$, $ABEFG$}, {F, $ABCDEF$, $CDEG$, ABG}, and {G, $ABCDEG$, $CDEF$, ABF}.

b. 1: (1), *aef, beg, abcd, abfg, cdfg, acdeg, bcdef*; 2: *ab, cd, fg, aeg, bef, acdef, bcdeg, abcdfg*; 3: *de, acg, adf, bcf, bdg, abce, cefg, abdefg*; 4: *ce, acf, adg, bcg, bdf, abde, defg, abcefg*.

49.

		A	B	C	D	E	AB	AC	AD	AE	BC	BD	BE	CD	CE	DE
a	70.4	+	−	−	−	−	−	−	−	−	+	+	+	+	+	+
b	72.1	−	+	−	−	−	−	+	+	+	−	−	−	+	+	+
c	70.4	−	−	+	−	−	+	−	+	+	−	+	+	−	−	+
abc	73.8	+	+	+	−	−	+	+	−	−	+	−	−	−	−	+
d	67.4	−	−	−	+	−	+	+	−	+	+	−	+	−	+	−
abd	67.0	+	+	−	+	−	+	−	+	−	−	+	−	+	−	−
acd	66.6	+	−	+	+	−	−	+	+	−	−	−	+	+	−	−
bcd	66.8	−	+	+	+	−	−	−	−	+	+	+	−	+	−	−
e	68.0	−	−	−	−	+	+	+	+	−	+	+	−	+	−	−
abe	67.8	+	+	−	−	+	+	−	−	+	−	−	+	+	−	−
ace	67.5	+	−	+	−	+	−	+	−	+	−	+	−	−	+	−
bce	70.3	−	+	+	−	+	−	−	+	−	+	−	+	−	+	−
ade	64.0	+	−	−	+	+	−	−	+	+	+	−	−	−	−	+
bde	67.9	−	+	−	+	+	−	+	−	−	−	+	+	−	−	+
cde	65.9	−	−	+	+	+	+	−	−	−	−	−	−	+	+	+
abcde	68.0	+	+	+	+	+	+	+	+	+	+	+	+	+	+	+

Thus $SSA = \dfrac{(70.4 - 72.1 - 70.4 + ... + 68.0)^2}{16} = 2.250$, $SSB = 7.840$, $SSC = .360$, $SSD = 52.563$, $SSE = 10.240$, $SSAB = 1.563$, $SSAC = 7.563$, $SSAD = .090$, $SSAE = 4.203$, $SSBC = 2.103$, $SSBD = .010$, $SSBE = .123$, $SSCD = .010$, $SSCE = .063$, $SSDE = 4.840$, Error SS = sum of two factor SS's = 20.568, Error MS = 2.057, $F_{.01,1,10} = 10.04$, so only the D main effect is significant.

Supplementary Exercises

51.

Source	Df	SS	MS	f
A	1	322.667	322.667	980.38
B	3	35.623	11.874	36.08
AB	3	8.557	2.852	8.67
Error	16	5.266	.329	
Total	23	372.113		

We first test the null hypothesis of no interactions ($H_0 : \gamma_{ij} = 0$ for all i, j). H_0 will be rejected at level .05

if $f_{AB} = \dfrac{MSAB}{MSE} \geq F_{.05,3,16} = 3.24$. Because $8.67 \geq 3.24$, H_0 is rejected. Because we have concluded

that interaction is present, tests for main effects are not appropriate.

53. Let A = spray volume, B = belt speed, C = brand.

Condition	Total	1	2	Contrast	$SS = \frac{(contrast)^2}{16}$
(1)	76	129	289	592	21,904.00
A	53	160	303	22	30.25
B	62	143	13	48	144.00
AB	98	160	9	134	1122.25
C	88	−23	31	14	12.25
AC	55	36	17	−4	1.00
BC	59	−33	59	−14	12.25
ABC	101	42	75	16	16.00

The ANOVA table is as follows:

Effect	Df	MS	f
A	1	30.25	6.72
B	1	144.00	32.00
AB	1	1122.25	249.39
C	1	12.25	2.72
AC	1	1.00	.22
BC	1	12.25	2.72
ABC	1	16.00	3.56
Error	8	4.50	
Total	15		

$F_{.05,1,8} = 5.32$, so all of the main effects are significant at level .05, but none of the interactions are significant.

55.

a.

Effect	%Iron	1	2	3	Effect Contrast	SS
	7	18	37	174	684	
A	11	19	137	510	144	1296
B	7	62	169	50	36	81
AB	12	75	341	94	0	0
C	21	79	9	14	272	4624
AC	41	90	41	22	32	64
BC	27	165	47	2	12	9
ABC	48	176	47	–2	–4	1
D	28	4	1	100	336	7056
AD	51	5	13	172	44	121
BD	33	20	11	32	8	4
ABD	57	21	11	0	0	0
CD	70	23	1	12	72	324
ACD	95	24	1	0	–32	64
BCD	77	25	1	0	–12	9
ABCD	99	22	–3	–4	–4	1

We use $estimate = \dfrac{contrast}{2^p}$ when $n = 1$ to get $\hat{\alpha}_1 = \dfrac{144}{2^4} = \dfrac{144}{16} = 9.00$, $\hat{\beta}_1 = \dfrac{36}{16} = 2.25$,

$\hat{\delta}_1 = \dfrac{272}{16} = 17.00$, $\hat{\gamma}_1 = \dfrac{336}{16} = 21.00$. Similarly, $\left(\widehat{\alpha\beta}\right)_{11} = 0$, $\left(\widehat{\alpha\delta}\right)_{11} = 2.00$,

$\left(\widehat{\alpha\gamma}\right)_{11} = 2.75$, $\left(\widehat{\beta\delta}\right)_{11} = .75$, $\left(\widehat{\beta\gamma}\right)_{11} = .50$, and $\left(\widehat{\delta\gamma}\right)_{11} = 4.50$.

b. The plot suggests main effects *A, C,* and *D* are quite important, and perhaps the interaction *CD* as well. In fact, pooling the 4 three-factor interaction SS's and the four-factor interaction SS to obtain an SSE based on 5 df and then constructing an ANOVA table suggests that these are the most important effects.

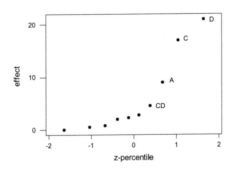

57. The ANOVA table is:

Source	df	SS	MS	f	$F_{.01}$
A	2	67553	33777	11.37	5.49
B	2	72361	36181	12.18	5.49
C	2	442111	221056	74.43	5.49
AB	4	9696	2424	0.82	4.11
AC	4	6213	1553	0.52	4.11
BC	4	34928	8732	2.94	4.11
ABC	8	33487	4186	1.41	3.26
Error	27	80192	2970		
Total	53	746542			

All three main effects are statistically significant at the 1% level, but no interaction terms are statistically significant at that level.

59. Based on the *P*-values in the ANOVA table, statistically significant factors at the level .01 are adhesive type and cure time. The conductor material does not have a statistically significant effect on bond strength. There are no significant interactions.

61. $SSA = \sum_i \sum_j \left(\bar{X}_{i...} - \bar{X}_{....}\right)^2 = \frac{1}{N}\sum X_{i...}^2 - \frac{X_{....}^2}{N}$, with similar expressions for SSB, SSC, and SSD, each having

$N-1$ df.

$SST = \sum_i \sum_j \left(X_{ij(kl)} - \bar{X}_{....}\right)^2 = \sum_i \sum_j X_{ij(kl)}^2 - \frac{X_{....}^2}{N}$ with $N^2 - 1$ df, leaving $N^2 - 1 - 4(N-1)$ df for error.

	1	2	3	4	5	Σx^2
$x_{i...}$:	482	446	464	468	434	1,053,916
$x_{.j..}$:	470	451	440	482	451	1,053,626
$x_{..k.}$:	372	429	484	528	481	1,066,826
$x_{...l}$:	340	417	466	537	534	1,080,170

Also, $\Sigma\Sigma x_{ij(kl)}^2 = 220{,}378$, $x_{....} = 2294$, and CF = 210,497.44.

Source	df	SS	MS	f	$F_{.05}$
A	4	285.76	71.44	.594	3.84
B	4	227.76	56.94	.473	3.84
C	4	2867.76	716.94	5.958*	3.84
D	4	5536.56	1384.14	11.502*	3.84
Error	8	962.72	120.34		
Total	24				

H_{0A} and H_{0B} cannot be rejected, while H_{0C} and H_{0D} are rejected.

CHAPTER 12

Section 12.1

1.

    a.   Stem and Leaf display of temp:

```
17|0
17|23           stem = tens
17|445          leaf = ones
17|67
17|
18|0000011
18|2222
18|445
18|6
18|8
```

180 appears to be a typical value for this data. The distribution is reasonably symmetric in appearance and somewhat bell-shaped. The variation in the data is fairly small since the range of values (188 – 170 = 18) is fairly small compared to the typical value of 180.

```
0|889
1|0000          stem = ones
1|3             leaf = tenths
1|4444
1|66
1|8889
2|11
2|
2|5
2|6
2|
3|00
```

For the ratio data, a typical value is around 1.6 and the distribution appears to be positively skewed. The variation in the data is large since the range of the data (3.08 - .84 = 2.24) is very large compared to the typical value of 1.6. The two largest values could be outliers.

    b.   The efficiency ratio is not uniquely determined by temperature since there are several instances in the data of equal temperatures associated with different efficiency ratios. For example, the five observations with temperatures of 180 each have different efficiency ratios.

162

c. A scatter plot of the data appears below. The points exhibit quite a bit of variation and do not appear to fall close to any straight line or simple curve.

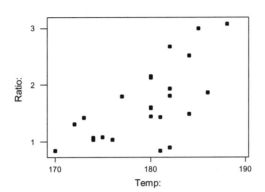

3. A scatter plot of the data appears below. The points fall very close to a straight line with an intercept of approximately 0 and a slope of about 1. This suggests that the two methods are producing substantially the same concentration measurements.

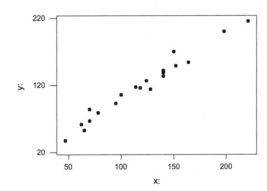

163

5.

    a. The scatter plot with axes intersecting at (0,0) is shown below.

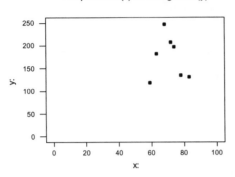

Temperature (x) vs Elongation (y)

    b. The scatter plot with axes intersecting at (55, 100) is shown below.

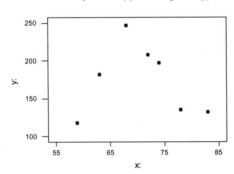

Temperature (x) vs Elongation (y)

    c. A parabola appears to provide a good fit to both graphs.

7.

    a. $\mu_{Y\cdot2500} = 1800 + 1.3(2500) = 5050$

    b. expected change = slope = $\beta_1 = 1.3$

    c. expected change = $100\beta_1 = 130$

    d. expected change = $-100\beta_1 = -130$

9.

a. β_1 = expected change in flow rate (y) associated with a one inch increase in pressure drop (x) = .095.

b. We expect flow rate to decrease by $5\beta_1 = .475$.

c. $\mu_{Y \cdot 10} = -.12 + .095(10) = .83$, and $\mu_{Y \cdot 15} = -.12 + .095(15) = 1.305$.

d. $P(Y > .835) = P\left(Z > \dfrac{.835 - .830}{.025}\right) = P(Z > .20) = .4207$

$P(Y > .840) = P\left(Z > \dfrac{.840 - .830}{.025}\right) = P(Z > .40) = .3446$

e. Let Y_1 and Y_2 denote pressure drops for flow rates of 10 and 11, respectively. Then $\mu_{Y \cdot 11} = .925$, so $Y_1 - Y_2$ has expected value .830 - .925 = -.095, and s.d. $\sqrt{(.025)^2 + (.025)^2} = .035355$. Thus

$P(Y_1 > Y_2) = P(Y_1 - Y_2 > 0) = P\left(z > \dfrac{+.095}{.035355}\right) = P(Z > 2.69) = .0036$

11.

a. β_1 = expected change for a one degree increase = -.01, and $10\beta_1 = -.1$ is the expected change for a 10 degree increase.

b. $\mu_{Y \cdot 200} = 5.00 - .01(200) = 3$, and $\mu_{Y \cdot 250} = 2.5$.

c. The probability that the first observation is between 2.4 and 2.6 is

$P(2.4 \le Y \le 2.6) = P\left(\dfrac{2.4 - 2.5}{.075} \le Z \le \dfrac{2.6 - 2.5}{.075}\right) = P(-1.33 \le Z \le 1.33) = .8164$. The

probability that any particular one of the other four observations is between 2.4 and 2.6 is also .8164, so the probability that all five are between 2.4 and 2.6 is $(.8164)^5 = .3627$.

d. Let Y_1 and Y_2 denote the times at the higher and lower temperatures, respectively. Then $Y_1 - Y_2$ has expected value $5.00 - .01(x+1) - (5.00 - .01x) = -.01$. The standard deviation of $Y_1 - Y_2$ is

$\sqrt{(.075)^2 + (.075)^2} = .10607$. Thus

$P(Y_1 - Y_2 > 0) = P\left(z > \dfrac{-(-.01)}{.10607}\right) = P(Z > .09) = .4641$.

Section 12.2

13. For this data, $n = 4$, $\Sigma x_i = 200$, $\Sigma y_i = 5.37$, $\Sigma x_i^2 = 12.000$, $\Sigma y_i^2 = 9.3501$, $\Sigma x_i y_i = 333$.

$$S_{xx} = 12{,}000 - \frac{(200)^2}{4} = 2000, \ S_{yy} = 9.3501 - \frac{(5.37)^2}{4} = 2.140875, \text{ and}$$

$$S_{xy} = 333 - \frac{(200)(5.37)}{4} = 64.5. \ \hat{\beta}_1 = \frac{S_{xy}}{S_{xx}} = \frac{64.5}{2000} = .03225 \text{ and}$$

$$\hat{\beta}_0 = \frac{5.37}{4} - (.03225)\frac{200}{4} = -.27000.$$

$$SSE = S_{yy} - \hat{\beta}_1 S_{xy} = 2.14085 - (.03225)(64.5) = .060750.$$

$$r^2 = 1 - \frac{SSE}{SST} = 1 - \frac{.060750}{2.14085} = .972.$$ This is a very high value of r^2, which confirms the authors' claim that there is a strong linear relationship between the two variables.

15.

a. The following stem and leaf display shows that: a typical value for this data is a number in the low 40's. There is some positive skew in the data. There are some potential outliers (79.5 and 80.0), and there is a reasonably large amount of variation in the data (e.g., the spread 80.0-29.8 = 50.2 is large compared with the typical values in the low 40's).

```
2 9
3 33          stem = tens
3 5566677889  leaf = ones
4 1223
4 56689
5 1
5
6 2
6 9
7
7 9
8 0
```

b. No, the strength values are not uniquely determined by the MoE values. For example, note that the two pairs of observations having strength values of 42.8 have different MoE values.

c. The least squares line is $\hat{y} = 3.2925 + .10748x$. For a beam whose modulus of elasticity is $x = 40$, the predicted strength would be $\hat{y} = 3.2925 + .10748(40) = 7.59$. The value $x = 100$ is far beyond the range of the x values in the data, so it would be dangerous (i.e., potentially misleading) to extrapolate the linear relationship that far.

d. From the output, SSE = 18.736, SST = 71.605, and the coefficient of determination is $r^2 = .738$ (or 73.8%). The r^2 value is large, which suggests that the linear relationship is a useful approximation to the true relationship between these two variables.

17.

a. From software, the equation of the least squares line is $\hat{y} = 118.91 - .905x$. The accompanying fitted line plot shows a very strong, linear association between unit weight and porosity. So, yes, we anticipate the linear model will explain a great deal of the variation in y.

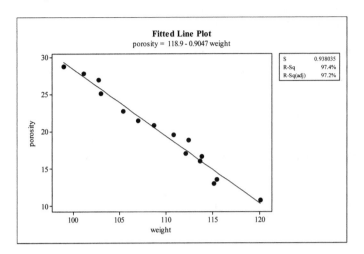

b. The slope of the line is $b_1 = -.905$. A one-pcf increase in the unit weight of a concrete specimen is associated with a .905 percentage point decrease in the specimen's predicted porosity. (Note: slope is not ordinarily a percent decrease, but the units on porosity, y, are percentage points.)

c. When $x = 135$, the predicted porosity is $\hat{y} = 118.91 - .905(135) = -3.265$. That is, we get a negative prediction for y, but in actuality y cannot be negative! This is an example of the perils of extrapolation; notice that $x = 135$ is outside the scope of the data.

d. The first observation is (99.0, 28.8). So, the actual value of y is 28.8, while the predicted value of y is $118.91 - .905(99.0) = 29.315$. The residual for the first observation is $e = y - \hat{y} = 28.8 - 29.315$ $= -.515 \approx -.52$.
Similarly, for the second observation we have $\hat{y} = 27.41$ and $e = 27.9 - 27.41 = .49$.

e. From software and the data provided, a point estimate of σ is $s = .938$. This represents the "typical" size of a deviation from the least squares line. More precisely, predictions from the least squares line are "typically" $\pm .938\%$ off from the actual porosity percentage.

f. From software, $r^2 = 97.4\%$ or .974, the proportion of observed variation in porosity that can be attributed to the approximate linear relationship between unit weight and porosity.

19. $n = 14$, $\Sigma x_i = 3300$, $\Sigma y_i = 5010$, $\Sigma x_i^2 = 913,750$, $\Sigma y_i^2 = 2,207,100$, $\Sigma x_i y_i = 1,413,500$

a. $\hat{\beta}_1 = \dfrac{3,256,000}{1,902,500} = 1.71143233$, $\hat{\beta}_0 = -45.55190543$, so we use the equation $y = -45.5519 + 1.7114x$.

b. $\hat{\mu}_{Y \cdot 225} = -45.5519 + 1.7114(225) = 339.51$.

c. Estimated expected change $= -50\hat{\beta}_1 = -85.57$

d. No, the value 500 is outside the range of x values for which observations were available (the danger of extrapolation).

21.

   a. Yes – a scatter plot of the data shows a strong, linear pattern, and $r^2 = 98.5\%$.

   b. From the output, the estimated regression line is $\hat{y} = 321.878 + 156.711x$, where x = absorbance and y = resistance angle. For x = .300, $\hat{y} = 321.878 + 156.711(.300) = 368.89$.

   c. The estimated regression line serves as an estimate both for a single y at a given x-value and for the true average μ_y at a given x-value. Hence, our estimate for μ_y when x = .300 is also 368.89.

23.

   a. Using the given y_i's and the formula $\hat{y}_i = -45.5519 + 1.7114x_i$,

$$SSE = (150 - 125.6)^2 + \dots + (670 - 639.0)^2 = 16,213.64 .$$ The computation formula gives
$$SSE = 2,207,100 - (-45.55190543)(5010) - (1.71143233)(1,413,500) = 16,205.45$$

   b. $SST = 2,207,100 - \dfrac{(5010)^2}{14} = 414,235.71$ so $r^2 = 1 - \dfrac{16,205.45}{414,235.71} = .961$.

25. Substitution of $\hat{\beta}_0 = \dfrac{\Sigma y_i - \hat{\beta}_1 \Sigma x_i}{n}$ and $\hat{\beta}_1$ for b_0 and b_1 on the left hand side of the normal equations

yields $\dfrac{n\left(\Sigma y_i - \hat{\beta}_1 \Sigma x_i\right)}{n} + \hat{\beta}_1 \Sigma x_i = \Sigma y_i$ from the first equation and

$$\dfrac{\Sigma x_i\left(\Sigma y_i - \hat{\beta}_1 \Sigma x_i\right)}{n} + \hat{\beta}_1 \Sigma x_i^2 = \dfrac{\Sigma x_i \Sigma y_i}{n} + \dfrac{\hat{\beta}_1\left(n\Sigma x_i^2 - (\Sigma x_i)^2\right)}{n}$$

$$\dfrac{\Sigma x_i \Sigma y_i}{n} + \dfrac{n\Sigma x_i y_i}{n} - \dfrac{\Sigma x_i \Sigma y_i}{n} = \Sigma x_i y_i$$ from the second equation.

27. We wish to find b_1 to minimize $f(b_1) = \Sigma(y_i - b_1 x_i)^2$. Equating $f'(b_1)$ to 0 yields $2\Sigma(y_i - b_1 x_i)(-x_i) = 0$

so $\Sigma x_i y_i = b_1 \Sigma x_i^2$ and $b_1 = \dfrac{\Sigma x_i y_i}{\Sigma x_i^2}$. The least squares estimator of $\hat{\beta}_1$ is thus $\hat{\beta}_1 = \dfrac{\Sigma x_i Y_i}{\Sigma x_i^2}$.

29. For data set #1, $r^2 = .43$ and $\hat{\sigma} = s = 4.03$; whereas these quantities are .99 and 4.03 for #2, and .99 and 1.90 for #3. In general, one hopes for both large r^2 (large % of variation explained) and small s (indicating that observations don't deviate much from the estimated line). Simple linear regression would thus seem to be most effective in the third situation.

Section 12.3

31.

   a. Software (Minitab) output from least squares regression on this data appears below. From the output, we see that $r^2 = 89.26\%$ or .8926, meaning 89.26% of the observed variation in threshold stress (y) can be attributed to the (approximate) linear model relationship with yield strength (x).

```
Regression Equation

y  =  211.655 - 0.175635 x

Coefficients

Term          Coef  SE Coef         T       P         95% CI
Constant   211.655  15.0622   14.0521   0.000  (178.503, 244.807)
x           -0.176   0.0184   -9.5618   0.000  ( -0.216,   -0.135)

Summary of Model

S = 6.80578      R-Sq = 89.26%      R-Sq(adj) = 88.28%
```

   b. From the software output, $\hat{\beta}_1 = -0.176$ and $s_{\hat{\beta}_1} = 0.0184$. Alternatively, the residual standard deviation is $s = 6.80578$, and the sum of square deviations of the x-values can be calculated to equal $S_{xx} = \sum (x_i - \bar{x})^2 = 138095$. From these, $s_{\hat{\beta}_1} = \dfrac{s}{\sqrt{S_{xx}}} = .0183$ (due to some slight rounding error).

   c. From the software output, a 95% CI for β_1 is (–0.216, –0.135). This is a fairly narrow interval, so β_1 has indeed been precisely estimated. Alternatively, with $n = 13$ we may construct a 95% CI for β_1 as $\hat{\beta}_1 \pm t_{.025,12} s_{\hat{\beta}_1} = -0.176 \pm 2.179(.0184) = (-0.216, -0.136)$.

33.

   a. Error df $= n - 2 = 25$, $t_{.025,25} = 2.060$, and so the desired confidence interval is

   $$\hat{\beta}_1 \pm t_{.025,25} \cdot s_{\hat{\beta}_1} = .10748 \pm (2.060)(.01280) = (.081, .134).$$ We are 95% confident that the true average change in strength associated with a 1 GPa increase in modulus of elasticity is between .081 MPa and .134 MPa.

   b. We wish to test $H_0 : \beta_1 \leq .1$ vs. $H_a : \beta_1 > .1$. The calculated t statistic is

   $$t = \frac{\hat{\beta}_1 - .1}{s_{\hat{\beta}_1}} = \frac{.10748 - .1}{.01280} = .58,$$ which yields a P-value of .277 at 25 df. Thus, we fail to reject

   H_0; i.e., there is not enough evidence to contradict the prior belief.

35.

a. We want a 95% CI for β_1: $\hat{\beta}_1 \pm t_{.025,15} \cdot s_{\hat{\beta}_1}$. First, we need our point estimate, $\hat{\beta}_1$. Using the given

summary statistics, $S_{xx} = 3056.69 - \dfrac{(222.1)^2}{17} = 155.019$, $S_{xy} = 2759.6 - \dfrac{(222.1)(193)}{17} = 238.112$,

and $\hat{\beta}_1 = \dfrac{S_{xy}}{S_{xx}} = \dfrac{238.112}{115.019} = 1.536$. We need $\hat{\beta}_0 = \dfrac{193 - (1.536)(222.1)}{17} = -8.715$ to calculate the SSE:

$SSE = 2975 - (-8.715)(193) - (1.536)(2759.6) = 418.2494$. Then $s = \sqrt{\dfrac{418.2494}{15}} = 5.28$ and

$s_{\hat{\beta}_1} = \dfrac{5.28}{\sqrt{155.019}} = .424$. With $t_{.025,15} = 2.131$, our CI is $1.536 \pm 2.131 \cdot (.424) = (.632,$

2.440). With 95% confidence, we estimate that the change in reported nausea percentage for every one-unit change in motion sickness dose is between .632 and 2.440.

b. We test the hypotheses $H_o : \beta_1 = 0$ vs $H_a : \beta_1 \neq 0$, and the test statistic is $t = \dfrac{1.536}{.424} = 3.6226$.

With df = 15, the two-tailed p-value = 2P(t > 3.6226) = 2(.001) = .002. With a p-value of .002, we would reject the null hypothesis at most reasonable significance levels. This suggests that there is a useful linear relationship between motion sickness dose and reported nausea.

c. No. A regression model is only useful for estimating values of nausea % when using dosages between 6.0 and 17.6 – the range of values sampled.

d. Removing the point (6.0, 2.50), the new summary stats are: $n = 16, \Sigma x_i = 216.1, \Sigma y_i = 191.5$,

$\Sigma x_i^2 = 3020.69, \Sigma y_i^2 = 2968.75, \Sigma x_i y_i = 2744.6$, and then $\hat{\beta}_1 = 1.561, \hat{\beta}_0 = -9.118$,

SSE = 430.5264, $s = 5.55$, $s_{\hat{\beta}_1} = .551$, and the new CI is $1.561 \pm 2.145 \cdot (.551)$, or (.379,

2.743). The interval is a little wider. But removing the one observation did not change it that much. The observation does not seem to be exerting undue influence.

37.

a. Let μ_d = the true mean difference in velocity between the two planes. We have 23 pairs of data that we will use to test H$_0$: $\mu_d = 0$ v. H$_a$: $\mu_d \neq 0$. From software, $\bar{x}_d = 0.2913$ with $s_d = 0.1748$, and so $t =$

$\dfrac{0.2913 - 0}{0.1748} \approx 8$, which has a two-sided P-value of 0.000 at 22 df. Hence, we strongly reject the null

hypothesis and conclude there is a statistically significant difference in true average velocity in the two planes. [*Note:* A normal probability plot of the differences shows one mild outlier, so we have slight concern about the results of the *t* procedure.]

b. Let β_1 denote the true slope for the linear relationship between Level $--$ velocity and Level – velocity.

We wish to test H$_0$: $\beta_1 = 1$ v. H$_a$: $\beta_1 < 1$. Using the relevant numbers provided, $t = \dfrac{b_1 - 1}{s(b_1)} = \dfrac{0.65393 - 1}{0.05947}$

$= -5.8$, which has a one-sided P-value at 23–2 = 21 df of P($t < -5.8$) ≈ 0. Hence, we strongly reject the null hypothesis and conclude the same as the authors; i.e., the true slope of this regression relationship is significantly less than 1.

39. SSE = 124,039.58 − (72.958547)(1574.8) − (.04103377)(222657.88) = 7.9679, and SST = 39.828

Source	df	SS	MS	f
Regr	1	31.860	31.860	18.0
Error	18	7.968	1.77	
Total	19	39.828		

Let's use $\alpha = .001$. Then $F_{.001,1,18} = 15.38 < 18.0$, so $H_o : \beta_1 = 0$ is rejected and the model is judged useful. $s = \sqrt{1.77} = 1.33041347$, $S_{xx} = 18,921.8295$, so

$$t = \frac{.04103377}{1.33041347 / \sqrt{18,921.8295}} = 4.2426, \text{ and } t^2 = (4.2426)^2 = 18.0 = f.$$

41.

a. Under the regression model, $E(Y_i) = \beta_0 + \beta_1 x_i$ and, hence, $E(\overline{Y}) = \beta_0 + \beta_1 \overline{x}$. Therefore,

$$E(Y_i - \overline{Y}) = \beta_1 (x_i - \overline{x}), \text{ and } E(\hat{\beta}_1) = E\left[\frac{\sum (x_i - \overline{x})(Y_i - \overline{Y})}{\sum (x_i - \overline{x})^2}\right] = \frac{\sum (x_i - \overline{x})E[Y_i - \overline{Y}]}{\sum (x_i - \overline{x})^2}$$

$$= \frac{\sum (x_i - \overline{x})\beta_1 (x_i - \overline{x})}{\sum (x_i - \overline{x})^2} = \beta_1 \frac{\sum (x_i - \overline{x})^2}{\sum (x_i - \overline{x})^2} = \beta_1.$$

b. With $c = \Sigma (x_i - \overline{x})^2$, $\hat{\beta}_1 = \frac{1}{c}\Sigma (x_i - \overline{x})(Y_i - \overline{Y}) = \frac{1}{c}\Sigma (x_i - \overline{x})Y_i$ (since

$\Sigma (x_i - \overline{x})\overline{Y} = \overline{Y}\Sigma (x_i - \overline{x}) = \overline{Y} \cdot 0 = 0$), so $V(\hat{\beta}_1) = \frac{1}{c^2}\Sigma (x_i - \overline{x})^2 Var(Y_i)$

$$= \frac{1}{c^2}\Sigma (x_i - \overline{x})^2 \cdot \sigma^2 = \frac{\sigma^2}{\Sigma (x_i - \overline{x})^2} = \frac{\sigma^2}{\Sigma x_i^2 - (\Sigma x_i)^2 / n}, \text{ as desired.}$$

43. The numerator of d is $|1 - 2| = 1$, and the denominator is $\frac{4\sqrt{14}}{\sqrt{324.40}} = .831$, so $d = \frac{1}{.831} = 1.20$. The

approximate power curve is for n − 2 df = 13, and β is read from Table A.17 as approximately .1.

Section 12.4

45.

    a. We wish to find a 90% CI for $\mu_{y\cdot125}$: $\hat{y}_{125} = 78.088$, $t_{.05,18} = 1.734$, and

$$s_{\hat{y}} = s\sqrt{\frac{1}{20} + \frac{(125 - 140.895)^2}{18,921.8295}} = .1674 .\text{Putting it together, we get}$$

$$78.088 \pm 1.734(.1674) = (77.797, 78.378)$$

    b. We want a 90% PI: Only the standard error changes: $s_{\hat{y}} = s\sqrt{1 + \frac{1}{20} + \frac{(125 - 140.895)^2}{18,921.8295}} = .6860$,

        so the PI is $78.088 \pm 1.734(.6860) = (76.898, 79.277)$

    c. Because the x^* of 115 is farther away from $\bar{x}$ than the previous value, the term $(x^* - \bar{x})^2$ will be larger, making the standard error larger, and thus the width of the interval is wider.

    d. We would be testing to see if the filtration rate were 125 kg-DS/m/h, would the average moisture content of the compressed pellets be less than 80%. The test statistic is $t = \dfrac{78.088 - 80}{.1674} = -11.42$, and with 18 df the p-value is P(t<-11.42) ≈ 0.00. We would reject H_o. There is significant evidence to prove that the true average moisture content when filtration rate is 125 is less than 80%.

47.

    a. $\hat{y}_{(40)} = -1.128 + .82697(40) = 31.95$, $t_{.025,13} = 2.160$; a 95% PI for runoff is

$$31.95 \pm 2.160\sqrt{(5.24)^2 + (1.44)^2} = 31.95 \pm 11.74 = (20.21, 43.69).$$ No, the resulting interval is very wide, therefore the available information is not very precise.

    b. $\Sigma x = 798, \Sigma x^2 = 63,040$ which gives $S_{xx} = 20,586.4$, which in turn gives

$$s_{\hat{y}_{(50)}} = 5.24\sqrt{\frac{1}{15} + \frac{(50 - 53.20)^2}{20,586.4}} = 1.358,$$ so the PI for runoff when x = 50 is

$$40.22 \pm 2.160\sqrt{(5.24)^2 + (1.358)^2} = 40.22 \pm 11.69 = (28.53, 51.92).$$ The simultaneous prediction level for the two intervals is at least $100(1 - 2\alpha)\% = 90\%$.

49.   95% CI: (462.1, 597.7); midpoint = 529.9; $t_{.025,8} = 2.306$; $529.9 + (2.306)\left(\hat{s}_{\hat{\beta}_0 + \hat{\beta}_1(15)}\right) = 597.7$

$$\Rightarrow \hat{s}_{\hat{\beta}_0 + \hat{\beta}_1(15)} = 29.402 \Rightarrow 99\% \text{ CI} = 529.9 \pm (3.355)(29.402) = (431.3, 628.5)$$

51.

    a. 0.40 is closer to $\bar{x}$.

    b. $\hat{\beta}_0 + \hat{\beta}_1(0.40) \pm t_{\alpha/2, n-2} \cdot \left(\hat{s}_{\hat{\beta}_0 + \hat{\beta}_1(0.40)}\right)$ or $0.8104 \pm (2.101)(0.0311) = (0.745, 0.876)$

    c. $\hat{\beta}_0 + \hat{\beta}_1(1.20) \pm t_{\alpha/2, n-2} \cdot \sqrt{s^2 + s^2_{\hat{\beta}_0 + \hat{\beta}_1(1.20)}}$ or

    $0.2912 \pm (2.101) \cdot \sqrt{(0.1049)^2 + (0.0352)^2} = (.059, .523)$

53.   Choice **a** will be the smallest, with **d** being largest. **a** is less than **b** and **c** (obviously), and **b** and **c** are both smaller than **d**. Nothing can be said about the relationship between **b** and **c**.

55. $\hat{\beta}_0 + \hat{\beta}_1 x = \bar{Y} - \hat{\beta}_1 \bar{x} + \hat{\beta}_1 x = \bar{Y} + (x - \bar{x})\hat{\beta}_1 = \dfrac{1}{n}\sum Y_i + \dfrac{(x - \bar{x})\sum(x_i - \bar{x})Y_i}{S_{XX}} = \sum d_i Y_i$ where

$d_i = \dfrac{1}{n} + \dfrac{(x - \bar{x})(x_i - \bar{x})}{S_{XX}}$. Thus, $Var\left(\hat{\beta}_0 + \hat{\beta}_1 x\right) = \sum d_i^2 Var(Y_i) = \sigma^2 \sum d_i^2$

$= \sigma^2 \sum \left[\dfrac{1}{n^2} + 2\dfrac{(x - \bar{x})(x_i - \bar{x})}{nS_{xx}} + \dfrac{(x - \bar{x})^2 (x_i - \bar{x})^2}{nS_{XX}^2} \right]$

$= \sigma^2 \left[n\dfrac{1}{n^2} + 2\dfrac{(x - \bar{x})\sum(x_i - \bar{x})}{nS_{xx}} + \dfrac{(x - \bar{x})^2 \sum(x_i - \bar{x})^2}{S_{XX}^2} \right]$

$= \sigma^2 \left[\dfrac{1}{n} + 2\dfrac{(x - \bar{x}) \cdot 0}{nS_{xx}} + \dfrac{(x - \bar{x})^2 S_{XX}}{S_{XX}^2} \right] = \sigma^2 \left[\dfrac{1}{n} + \dfrac{(x - \bar{x})^2}{S_{XX}} \right]$

Section 12.5

57.   Most people acquire a license as soon as they become eligible. If, for example, the minimum age for obtaining a license is 16, then the time since acquiring a license, y, is usually related to age by the equation $y \approx x - 16$, which is the equation of a straight line. In other words, the majority of people in a sample will have y values that closely follow the line $y = x - 16$.

59.

    a. $S_{xx} = 251{,}970 - \dfrac{(1950)^2}{18} = 40{,}720$, $S_{yy} = 130.6074 - \dfrac{(47.92)^2}{18} = 3.033711$, and

    $S_{xy} = 5530.92 - \dfrac{(1950)(47.92)}{18} = 339.586667$, so $r = \dfrac{339.586667}{\sqrt{40{,}720}\sqrt{3.033711}} = .9662$.

    There is a very strong positive correlation between the two variables.

    b.   Because the association between the variables is positive, the specimen with the larger shear force will tend to have a larger percent dry fiber weight.

c. Changing the units of measurement on either (or both) variables will have no effect on the calculated value of r, because any change in units will affect both the numerator and denominator of r by exactly the same multiplicative constant.

d. $r^2 = (.966)^2 = .933$

e. $H_0 : \rho = 0$ vs $H_a : \rho > 0$. $t = \dfrac{r\sqrt{n-2}}{\sqrt{1-r^2}}$; Reject H_0 at level .01 if $t \geq t_{.01,16} = 2.583$.

$t = \dfrac{.966\sqrt{16}}{\sqrt{1-.966^2}} = 14.94 \geq 2.583$, so H_o should be rejected. The data indicates a positive linear relationship between the two variables.

61.

a. We are testing $H_0 : \rho = 0$ vs $H_a : \rho > 0$. $r = \dfrac{7377.704}{\sqrt{36.9839}\sqrt{2,628,930.359}} = .7482$, and

$t = \dfrac{.7482\sqrt{12}}{\sqrt{1-.7482^2}} = 3.9066$. We reject H_o since $t = 3.9066 \geq t_{.05,12} = 1.782$. There is evidence that a positive correlation exists between maximum lactate level and muscular endurance.

b. We are looking for r^2, the coefficient of determination. $r^2 = (.7482)^2 = .5598$. It is the same no matter which variable is the predictor.

63. $n = 6$, $\Sigma x_i = 111.71, \Sigma x_i^2 = 2,724.7643, \Sigma y_i = 2.9, \Sigma y_i^2 = 1.6572$, and $\Sigma x_i y_i = 63.915$.

$r = \dfrac{(6)(63.915)-(111.71)(2.9)}{\sqrt{(6)(2,724.7943)-(111.73)^2} \cdot \sqrt{(6)(1.6572)-(2.9)^2}} = .7729$. $H_0 : \rho = 0$ vs $H_a : \rho \neq 0$; Reject H_o

at level .05 if $|t| \geq t_{.025,4} = 2.776$. $t = \dfrac{(.7729)\sqrt{4}}{\sqrt{1-(.7729)^2}} = 2.436$. Fail to reject H_o. The data does not indicate

that the population correlation coefficient differs from 0. This result may seem surprising due to the relatively large size of r (.77), however, it can be attributed to a small sample size (6).

65.

a. From the summary statistics provided, a point estimate for the population correlation coefficient ρ is r

$= \dfrac{\sum(x_i - \bar{x})(y_i - \bar{y})}{\sqrt{\sum(x_i - \bar{x})^2 \sum(y_i - \bar{y})^2}} = \dfrac{44,185.87}{\sqrt{(64,732.83)(130,566.96)}} = .4806$.

b. The hypotheses are $H_0: \rho = 0$ versus $H_a: \rho \neq 0$. Assuming bivariate normality, the test statistic value is

$t = \dfrac{r\sqrt{n-2}}{\sqrt{1-r^2}} = \dfrac{.4806\sqrt{15-2}}{\sqrt{1-.4806^2}} = 1.98$. At df $= 15 - 2 = 13$, the two-tailed P-value for this t test is $2P(T_{13}$

$\geq 1.98) \approx 2P(T_{13} \geq 2.0) = 2(.033) = .066$. Hence, we fail to reject H_0 at the .01 level; there is not sufficient evidence to conclude that the population correlation coefficient between internal and external rotation velocity is not zero.

c. If we tested H_0: $\rho = 0$ versus H_a: $\rho > 0$, the one-sided *P*-value would be .033. We would still fail to reject H_0 at the .01 level, lacking sufficient evidence to conclude a positive true correlation coefficient. However, for a one-sided test at the .05 level, we would reject H_0 since *P*-value = .033 < .05. We have evidence at the .05 level that the true population correlation coefficient between internal and external rotation velocity is positive.

67.

    a. Because p-value = .00032 < α = .001, H_o should be rejected at this significance level.

    b. Not necessarily. For this n, the test statistic *t* has approximately a standard normal distribution when $H_0 : \rho_1 = 0$ is true, and a p-value of .00032 corresponds to $z = 3.60$ (or -3.60). Solving

$$3.60 = \frac{r\sqrt{498}}{\sqrt{1-r^2}}$$ for r yields r = .159. This r suggests only a weak linear relationship between x and y, one that would typically have little practical importance .

    c. $t = \dfrac{.022\sqrt{9998}}{\sqrt{1-.022^2}} = 2.20 \geq t_{.025,9998} = 1.96$, so H_0 is rejected in favor of H_a. The value $t = 2.20$ is statistically significant – it cannot be attributed just to sampling variability in the case $\rho = 0$. But with this *n*, r = .022 implies $\rho \approx .022$, which in turn shows an extremely weak linear relationship.

Supplementary Exercises

69. Use available software for all calculations.

    a. We want a confidence interval for β_1. From software, $b_1 = 0.987$ and $s(b_1) = 0.047$, so the corresponding 95% CI is $0.987 \pm t_{.025,17}(0.047) = 0.987 \pm 2.110(0.047) = (0.888, 1.086)$. We are 95% confident that the true average change in sale price associated with a one-foot increase in truss height is between $0.89 per square foot and $1.09 per square foot.

    b. Using software, a 95% CI for $\mu_{y.25}$ is (47.730, 49.172). We are 95% confident that the true average sale price for all warehouses with 25-foot truss height is between $47.73/ft^2 and $49.17/ft^2.

    c. Again using software, a 95% PI for Y when x = 25 is (45.378, 51.524). We are 95% confident that the sale price for a single warehouse with 25-foot truss height will be between $45.38/ft^2 and $51.52/ft^2.

    d. Since x = 25 is nearer the mean than x = 30, a PI at x = 30 would be wider.

    e. From software, r^2 = SSR/SST = 890.36/924.44 = .963. Hence, $r = \sqrt{.963} = .981$.

71.

a. The test statistic value is $t = \dfrac{\hat{\beta}_1 - 1}{s_{\hat{\beta}_1}}$, and H_0 will be rejected if either $t \geq t_{.025,11} = 2.201$ or $t \leq -2.201$.

With $\Sigma x_i = 243, \Sigma x_i^2 = 5965, \Sigma y_i = 241, \Sigma y_i^2 = 5731$ and $\Sigma x_i y_i = 5805$, $\hat{\beta}_1 = .913819$, $\hat{\beta}_0 = 1.457072$,

$SSE = 75.126$, $s = 2.613$, and $s_{\hat{\beta}_1} = .0693$, $t = \dfrac{.9138 - 1}{.0693} = -1.24$. Because -1.24 is neither ≤ -2.201

nor ≥ 2.201, H_0 cannot be rejected. It is plausible that $\beta_1 = 1$.

b. $r = \dfrac{16{,}902}{(136)(128.15)} = .970.$

73.

a. $r^2 = .5073$

b. $r = +\sqrt{r^2} = \sqrt{.5073} = .7122$ (positive because $\hat{\beta}_1$ is positive.)

c. We test $H_0 : \beta_1 = 0$ v. $H_a : \beta_1 \neq 0$.. The test statistic t = 3.93 gives p-value = .0013, which is < .01, the given level of significance, therefore we reject H_o and conclude that the model is useful.

d. We use a 95% CI for $\mu_{Y \cdot 50}$. $\hat{y}_{(50)} = .787218 + .007570(50) = 1.165718$, $t_{.025,15} = 2.131$, s =

"Root MSE" = .020308, so $s_{\hat{y}_{(50)}} = .20308\sqrt{\dfrac{1}{17} + \dfrac{17(50 - 42.33)^2}{17(41{,}575) - (719.60)^2}} = .051422$. The

interval is, then, $1.165718 \pm 2.131(.051422) = 1.165718 \pm .109581 = (1.056137, 1.275299)$

.

e. $\hat{y}_{(30)} = .787218 + .007570(30) = 1.0143$. The residual is $y - \hat{y} = .80 - 1.0143 = -.2143$.

75.

a. n = 9, $\Sigma x_i = 228, \Sigma x_i^2 = 5958, \Sigma y_i = 93.76, \Sigma y_i^2 = 982.2932$ and $\Sigma x_i y_i = 2348.15$, giving

$\hat{\beta}_1 = \dfrac{-243.93}{1638} = -.148919$, $\hat{\beta}_0 = 14.190392$, and the equation $\hat{y} = 14.19 - (.1489)x$.

b. β_1 is the expected increase in load associated with a one-day age increase (so a negative value of β_1 corresponds to a decrease). We wish to test $H_0 : \beta_1 = -.10$ v. $H_a : \beta_1 < -.10$ (the alternative

contradicts prior belief). H_o will be rejected at level .05 if $t = \dfrac{\hat{\beta}_1 - (-.10)}{s_{\hat{\beta}_1}} \leq -t_{.05,7} = -1.895$.

With SSE = 1.4862, s = .4608, and $s_{\hat{\beta}_1} = \dfrac{.4608}{\sqrt{182}} = .0342$. Thus $t = \dfrac{-.1489 + .1}{.0342} = -1.43$.

Because -1.43 is not ≤ -1.895, do not reject H_o. The data do not contradict this assertion.

c. $\Sigma x_i = 306, \Sigma x_i^2 = 7946$, so $\sum (x_i - \bar{x})^2 = 7946 - \dfrac{(306)^2}{12} = 143$ here, as contrasted with 182

for the given 9 x_i's. Even though the sample size for the proposed x values is larger, the original set of values is preferable.

d. $(t_{.025,7})(s)\sqrt{\dfrac{1}{9} + \dfrac{9(28 - 25.33)^2}{1638}} = (2.365)(.4608)(.3877) = .42$, and $\hat{\beta}_0 + \hat{\beta}_1(28) = 10.02$,

so the 95% CI is $10.02 \pm .42 = (9.60, 10.44)$.

77.

a.   The plot suggests a strong linear relationship between x and y.

b. $n = 9, \Sigma x_i = 1797, \Sigma x_i^2 = 4334.41, \Sigma y_i = 7.28, \Sigma y_i^2 = 7.4028$ and $\Sigma x_i y_i = 178.683$, so

$\hat{\beta}_1 = \dfrac{299.931}{6717.6} = .04464854$, $\hat{\beta}_0 = -.08259353$, and the equation of the estimated line is

$\hat{y} = -.08259 + .044649x$.

c. $SSE = 7.4028 - (-601281) - 7.977935 = .026146$, $SST = 7.4028 - \dfrac{(7.28)^2}{9} = 1.5141$, and

$r^2 = 1 - \dfrac{SSE}{SST} = .983$, so 93.8% of the observed variation can be explained by the model relationship.

d. $\hat{y}_4 = -.08259 - (.044649)(19.1) = .7702$, and $y_4 - \hat{y}_4 = .68 - .7702 = -.0902$.

e. $s = .06112$, and $s_{\hat{\beta}_1} = \dfrac{.06112}{\sqrt{746.4}} = .002237$, so the value of t for testing $H_0 : \beta_1 = 0$ vs

$H_0 : \beta_1 \neq 0$ is $t = \dfrac{.044649}{.002237} = 19.96$. From Table A.5, $t_{.0005,7} = 5.408$, so the 2-sided P-value is

less than 2(.005) = .001. There is strong evidence for a useful relationship.

f.   A 95% CI for β_1 is $.044649 \pm (2.365)(.002237) = .044649 \pm .005291 = (.0394, .0499)$.

g.   A 95% CI for $\beta_0 + \beta_1(20)$ is $.810 \pm (2.365)(.002237)(.3333356)$

$= .810 \pm .048 = (.762, .858)$

79. $SSE = \Sigma y^2 - \hat{\beta}_0 \Sigma y - \hat{\beta}_1 \Sigma xy$. Substituting $\hat{\beta}_0 = \dfrac{\Sigma y - \hat{\beta}_1 \Sigma x}{n}$, SSE becomes

$$SSE = \Sigma y^2 - \frac{\Sigma y \left(\Sigma y - \hat{\beta}_1 \Sigma x \right)}{n} - \hat{\beta}_1 \Sigma xy = \Sigma y^2 - \frac{\left(\Sigma y \right)^2}{n} + \frac{\hat{\beta}_1 \Sigma x \Sigma y}{n} - \hat{\beta}_1 \Sigma xy$$

$$= \left[\Sigma y^2 - \frac{\left(\Sigma y \right)^2}{n} \right] - \hat{\beta}_1 \left[\Sigma xy - \frac{\Sigma x \Sigma y}{n} \right] = S_{yy} - \hat{\beta}_1 S_{xy} , \text{ as desired.}$$

81.

a. With $s_{xx} = \sum (x_i - \bar{x})^2$, $s_{yy} = \sum (y_i - \bar{y})^2$, note that $\dfrac{s_y}{s_x} = \sqrt{\dfrac{s_{yy}}{s_{xx}}}$ (since the factor n-1

appears in both the numerator and denominator, so cancels). Thus

$$y = \hat{\beta}_0 + \hat{\beta}_1 x = \bar{y} + \hat{\beta}_1 (x - \bar{x}) = \bar{y} + \frac{s_{xy}}{s_{xx}} (x - \bar{x}) = \bar{y} + \sqrt{\frac{s_{yy}}{s_{xx}}} \cdot \frac{s_{xy}}{\sqrt{s_{xx} s_{yy}}} (x - \bar{x}) = \bar{y} + \frac{s_y}{s_x} \cdot r \cdot (x - \bar{x}),$$

as desired.

b. By .573 s.d.'s above, (above, since r < 0) or (since s_y = 4.3143) an amount 2.4721 above.

83. Using the notation of the previous exercise, $SST = s_{yy}$ and $SSE = s_{yy} - \hat{\beta}_1 s_{xy} = s_{yy} - \dfrac{s_{xy}^2}{s_{xx}}$, so

$$1 - \frac{SSE}{SST} = 1 - \frac{s_{yy} - \dfrac{s_{xy}^2}{s_{xx}}}{s_{yy}} = \frac{s_{xy}^2}{s_{xx} s_{yy}} = r^2 , \text{ as desired.}$$

85. Using Minitab, we create a scatterplot to see if a linear regression model is appropriate.

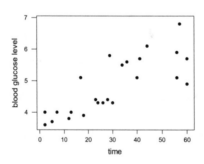

A linear model is reasonable; although it appears that the variance in y gets larger as x increases. The Minitab output follows:

```
The regression equation is
blood glucose level = 3.70 + 0.0379 time

Predictor          Coef         StDev           T          P
Constant         3.6965        0.2159       17.12      0.000
time            0.037895      0.006137        6.17      0.000

S = 0.5525      R-Sq = 63.4%      R-Sq(adj) = 61.7%

Analysis of Variance

Source             DF            SS          MS          F          P
Regression          1        11.638      11.638      38.12      0.000
Residual Error     22         6.716       0.305
Total              23        18.353
```

The coefficient of determination of 63.4% indicates that only a moderate percentage of the variation in y can be explained by the change in x. A test of model utility indicates that time is a significant predictor of blood glucose level. ($t = 6.17$, $p = 0.0$). A point estimate for blood glucose level when time = 30 minutes is 4.833%. We would expect the average blood glucose level at 30 minutes to be between 4.599 and 5.067, with 95% confidence.

87. For the second boiler, $n = 6$, $\Sigma x_i = 125$, $\Sigma y_i = 472.0$, $\Sigma x_i^2 = 3625$, $\Sigma y_i^2 = 37,140.82$, and

$\Sigma x_i y_i = 9749.5$, giving $\hat{\gamma}_1$ = estimated slope = $\dfrac{-503}{6125} = -.0821224$, $\hat{\gamma}_0 = 80.377551$,

$SSE_2 = 3.26827$, $SSx_2 = 1020.833$. For boiler #1, $n = 8$, $\hat{\beta}_1 = -.1333$, $SSE_1 = 8.733$, and

$SSx_1 = 1442.875$. Thus $\hat{\sigma}^2 = \dfrac{8.733 + 3.286}{10} = 1.2$, $\hat{\sigma} = 1.095$, and

$t = \dfrac{-.1333 + .0821}{1.095\sqrt{\frac{1}{1442.875} + \frac{1}{1020.833}}} = \dfrac{-.0512}{.0448} = -1.14$. $t_{.025,10} = 2.228$ and -1.14 is neither ≥ 2.228

nor ≤ -2.228, so H_o is not rejected. It is plausible that $\beta_1 = \gamma_1$.

CHAPTER 13

Section 13.1

1.

   a. $\bar{x} = 15$ and $\sum(x_j - \bar{x})^2 = 250$, so s.d. of $Y_i - \hat{Y}_i$ is $10\sqrt{1 - \dfrac{1}{5} - \dfrac{(x_i - 15)^2}{250}} = 6.32, 8.37,$
     8.94, 8.37, and 6.32 for i = 1, 2, 3, 4, 5.

   b. Now $\bar{x} = 20$ and $\sum(x_i - \bar{x})^2 = 1250$, giving standard deviations 7.87, 8.49, 8.83, 8.94, and
     2.83 for i = 1, 2, 3, 4, 5.

   c. The deviation from the estimated line is likely to be much smaller for the observation made in the
     experiment of **b** for x = 50 than for the experiment of **a** when x = 25. That is, the observation (50, Y)
     is more likely to fall close to the least squares line than is (25, Y).

3.

   a. This plot indicates there are no outliers, the variance of ε is reasonably constant, and the ε are normally
     distributed. A straight-line regression function is a reasonable choice for a model.

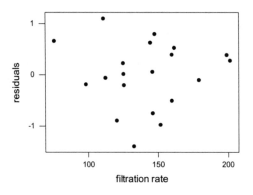

   b. We need $S_{xx} = \sum(x_i - \bar{x})^2 = 415{,}914.85 - \dfrac{(2817.9)^2}{20} = 18{,}886.8295$. Then each e_i^* can be

     calculated as follows: $e_i^* = \dfrac{e_i}{.4427\sqrt{1 + \dfrac{1}{20} + \dfrac{(x_i - 140.895)^2}{18{,}886.8295}}}$. The table below shows the values:

180

standardized residuals	e/e_i^*	standardized residuals	e/e_i^*
-0.31064	0.644053	0.6175	0.64218
-0.30593	0.614697	0.09062	0.64802
0.4791	0.578669	1.16776	0.565003
1.2307	0.647714	-1.50205	0.646461
-1.15021	0.648002	0.96313	0.648257
0.34881	0.643706	0.019	0.643881
-0.09872	0.633428	0.65644	0.584858
-1.39034	0.640683	-2.1562	0.647182
0.82185	0.640975	-0.79038	0.642113
-0.15998	0.621857	1.73943	0.631795

Notice that if $e_i^* \approx e/s$, then $e/e_i^* \approx s$. All of the e/e_i^*'s range between .57 and .65, which are close to s.

c. This plot looks very much the same as the one in part a.

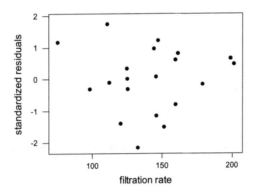

5.

a. 97.7% of the variation in ice thickness can be explained by the linear relationship between it and elapsed time. Based on this value, it is tempting to assume an approximately linear relationship; however, r^2 does <u>not</u> measure the aptness of the linear model.

b. The residual plot shows a curve in the data, suggesting a non-linear relationship exists. One observation (5.5, -3.14) is extreme.

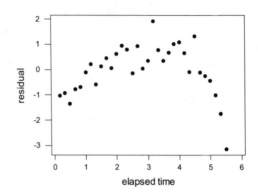

7.

a. From software and the data provided, the least squares line is $\hat{y} = 84.4 - 290x$. Also from software, the coefficient of determination is $r^2 = 77.6\%$ or .776.

Regression Analysis: y versus x

```
The regression equation is
y = 84.4 - 290 x

Predictor      Coef   SE Coef       T       P
Constant      84.38     11.64    7.25   0.000
x           -289.79     43.12   -6.72   0.000

S = 2.72669   R-Sq = 77.6%   R-Sq(adj) = 75.9%
```

b. The accompanying scatterplot exhibits substantial curvature, which suggests that a straight-line model is not actually a good fit.

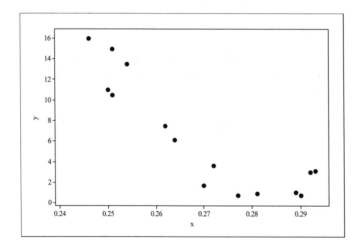

182

c. Fits, residuals, and standardized residuals were computed using software and the accompanying plot was created. The residual-versus-fit plot indicates very strong curvature but not a lack of constant variance. This implies that a linear model is inadequate, and a quadratic (parabolic) model relationship might be suitable for x and y.

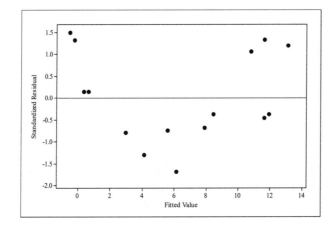

9. Both a scatter plot and residual plot (based on the simple linear regression model) for the first data set suggest that a simple linear regression model is reasonable, with no pattern or influential data points which would indicate that the model should be modified. However, scatter plots for the other three data sets reveal difficulties.

Scatter Plot for Data Set #1

Scatter Plot for Data Set #2

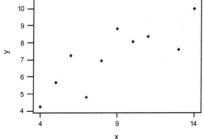

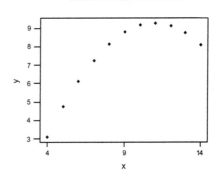

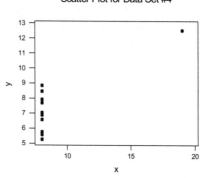

Scatter Plot for Data Set #3

Scatter Plot for Data Set #4

For data set #2, a quadratic function would clearly provide a much better fit. For data set #3, the relationship is perfectly linear except one outlier, which has obviously greatly influenced the fit even though its x value is not unusually large or small. One might investigate this observation to see whether it was mistyped and/or it merits deletion. For data set #4 it is clear that the slope of the least squares line has been determined entirely by the outlier, so this point is extremely influential. A linear model is completely inappropriate for data set #4; if anything, a modified logistic regression model (see Section 13.2) might be more appropriate.

11.

a. $Y_i - \hat{Y}_i = Y_i - \bar{Y} - \hat{\beta}_1(x_i - \bar{x}) = Y_i - \frac{1}{n}\sum_j Y_j - \frac{(x_i - \bar{x})\sum_j(x_j - \bar{x})Y_j}{\sum_j(x_j - \bar{x})^2} = \sum_j c_j Y_j$, where

$c_j = 1 - \frac{1}{n} - \frac{(x_i - \bar{x})^2}{n\Sigma(x_j - \bar{x})^2}$ for j = i and $c_j = 1 - \frac{1}{n} - \frac{(x_i - \bar{x})(x_j - \bar{x})}{\Sigma(x_j - \bar{x})^2}$ for $j \neq i$. Thus

$Var(Y_i - \hat{Y}_i) = \Sigma Var(c_j Y_j)$ (since the Y_j's are independent) $= \sigma^2 \Sigma c_j^2$ which, after some algebra, gives equation (13.2).

b. $\sigma^2 = Var(Y_i) = Var(\hat{Y}_i + (Y_i - \hat{Y}_i)) = Var(\hat{Y}_i) + Var(Y_i - \hat{Y}_i)$, so

$Var(Y_i - \hat{Y}_i) = \sigma^2 - Var(\hat{Y}_i) = \sigma^2 - \sigma^2\left[\frac{1}{n} + \frac{(x_i - \bar{x})^2}{\Sigma(x_j - \bar{x})^2}\right]$, which is exactly (13.2).

c. As x_i moves further from $\bar{x}$, $(x_i - \bar{x})^2$ grows larger, so $Var(\hat{Y}_i)$ increases (since $(x_i - \bar{x})^2$ has a positive sign in $Var(\hat{Y}_i)$), but $Var(Y_i - \hat{Y}_i)$ decreases (since $(x_i - \bar{x})^2$ has a negative sign).

13. The distribution of any particular standardized residual is also a t distribution with n – 2 d.f., since e_i^* is obtained by taking standard normal variable $\dfrac{\left(Y_i - \hat{Y}_i\right)}{\left(\sigma_{Y_i - \hat{Y}}\right)}$ and substituting the estimate of σ in the denominator (exactly as in the predicted value case). With E_i^* denoting the ith standardized residual as a random variable, when n = 25 E_i^* has a t distribution with 23 d.f. and $t_{.01,23} = 2.50$, so P(E_i^* outside (-2.50, 2.50)) = $P\left(E_i^* \geq 2.50\right) + P\left(E_i^* \leq -2.50\right) = .01 + .01 = .02$.

Section 13.2

15.

   a.

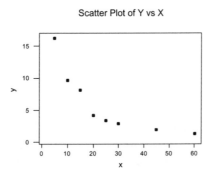

Scatter Plot of Y vs X

The plot has a curved pattern. A linear model would not be appropriate.

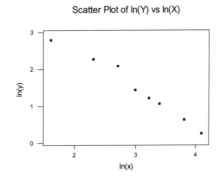

Scatter Plot of ln(Y) vs ln(X)

   b. In this plot we have a strong linear pattern.

   c. The linear pattern in **b** above would indicate that a transformed regression using the natural log of both x and y would be appropriate. The probabilistic model is then $y = \alpha x^{\beta} \cdot \varepsilon$ (the power function with an error term).

d. A regression of ln(y) on ln(x) yields the equation $\ln(y) = 4.6384 - 1.04920\ln(x)$. Using Minitab we can get a P.I. for y when x = 20 by first transforming the x value: $\ln(20) = 2.996$. The computer generated 95% P.I. for ln(y) when ln(x) = 2.996 is (1.1188,1.8712). We must now take the antilog to return to the original units of Y: $\left(e^{1.1188}, e^{1.8712}\right) = (3.06, 6.50)$.

e. A computer generated residual analysis:

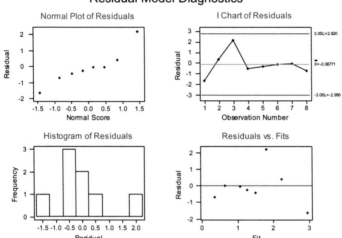

Looking at the residual vs. fits (bottom right), one standardized residual, corresponding to the third observation, is a bit large. There are only two positive standardized residuals, but two others are essentially 0. The patterns in the residual plot and the normal probability plot (upper left) are marginally acceptable.

17.

a. $\Sigma x_i' = 15.501$, $\Sigma y_i' = 13.352$, $\Sigma x_i'^2 = 20.228$, $\Sigma y_i'^2 = 16.572$, $\Sigma x_i'\ y_i' = 18.109$, from which $\hat{\beta}_1 = 1.254$ and $\hat{\beta}_0 = -.468$ so $\hat{\beta} = \hat{\beta}_1 = 1.254$ and $\hat{\alpha} = e^{-.468} = .626$.

b. The plots give strong support to this choice of model; in addition, $r^2 = .960$ for the transformed data.

c. SSE = .11536 (computer printout), s = .1024, and the estimated SD of $\hat{\beta}_1$ is .0775, so
$$t = \frac{1.25 - 1.33}{.0775} = -1.07.$$ Since −1.07 is not $\leq -t_{.05,11} = -1.796$, H_0 cannot be rejected in favor of H_a.

d. The claim that $\mu_{Y \cdot 5} = 2\mu_{Y \cdot 2.5}$ is equivalent to $\alpha \cdot 5^\beta = 2\alpha(2.5)^\beta$, or that $\beta = 1$. Thus we wish test $H_0 : \beta = 1$ vs. $H_a : \beta \neq 1$. With $t = \frac{1 - 1.33}{.0775} = -4.30 \leq -t_{.005,11} \leq -3.106$, H_0 is rejected at level .01.

19.

   a. No, there is definite curvature in the plot.

   b. With x = temperature and y = lifetime, a linear relationship between ln(lifetime) and 1/temperature implies a model $y = \exp(\alpha + \beta/x + \varepsilon)$. Let $x' = \dfrac{1}{temp}$ and $y' = \ln(lifetime)$. Plotting y' vs. x' gives a plot which has a pronounced linear appearance (and in fact $r^2 = .954$ for the straight line fit).

   c. $\Sigma x'_i = .082273$, $\Sigma y'_i = 123.64$, $\Sigma x'^2_i = .00037813$, $\Sigma y'^2_i = 879.88$, $\Sigma x'_i\, y'_i = .57295$, from which $\hat{\beta} = 3735.4485$ and $\hat{\alpha} = -10.2045$ (values read from computer output). With x = 220, $x' = .004545$ so $\hat{y}' = -10.2045 + 3735.4485(.004545) = 6.7748$ and thus $\hat{y} = e^{\hat{y}'} = 875.50$.

   d. For the transformed data, SSE = 1.39857, and $n_1 = n_2 = n_3 = 6$, $\bar{y}'_{1.} = 8.44695$, $\bar{y}'_{2.} = 6.83157$, $\bar{y}'_{3.} = 5.32891$, from which SSPE = 1.36594, SSLF = .02993,
$$f = \frac{.02993/1}{1.36594/15} = .33.$$
Comparing this to $F_{.01,1,15} = 8.68$, it is clear that H$_0$ cannot be rejected.

21.

   a. The suggested model is $Y = \beta_0 + \beta_1(x') + \varepsilon$ where $x' = \dfrac{10^4}{x}$. The summary quantities are $\Sigma x'_i = 159.01$, $\Sigma y_i = 121.50$, $\Sigma x'^2_i = 4058.8$, $\Sigma y^2_i = 1865.2$, $\Sigma x'_i\, y_i = 2281.6$, from which $\hat{\beta}_1 = -.1485$ and $\hat{\beta}_0 = 18.1391$, and the estimated regression function is
$$y = 18.1391 - \frac{1485}{x}.$$

   b. $x = 500 \Rightarrow \hat{y} = 18.1391 - \dfrac{1485}{500} = 15.17$.

23. $V(Y) = V(\alpha e^{\beta x} \cdot \varepsilon) = \left[\alpha e^{\beta x}\right]^2 \cdot V(\varepsilon) = \alpha^2 e^{2\beta x} \cdot \tau^2$ where we have set $V(\varepsilon) = \tau^2$. If $\beta > 0$, this is an increasing function of x so we expect more spread in y for large x than for small x, while the situation is reversed if $\beta < 0$. It is important to realize that a scatter plot of data generated from this model will not spread out uniformly about the exponential regression function throughout the range of x values; the spread will only be uniform on the transformed scale. Similar results hold for the multiplicative power model.

25.   First, the test statistic for the hypotheses H_0: $\beta_1 = 0$ versus H_a: $\beta_1 \neq 0$ is $z = -4.58$ with a corresponding P-value of .000, suggesting noise level has a highly statistically significant relationship with people's perception of the acceptability of the work environment. The negative value indicates that the likelihood of finding work environment acceptable <u>decreases</u> as the noise level increases (not surprisingly). We estimate that a 1 dBA increase in noise level decreases the <u>odds</u> of finding the work environment acceptable by a multiplicative factor of .70 (95% CI: .60 to .81).

The accompanying plot shows $\hat{\pi} = \dfrac{e^{b_0 + b_1 x}}{1 + e^{b_0 + b_1 x}} = \dfrac{e^{23.2 - .359x}}{1 + e^{23.2 - .359x}}$. Notice that the estimate probability of finding work environment acceptable decreases as noise level, x, increases.

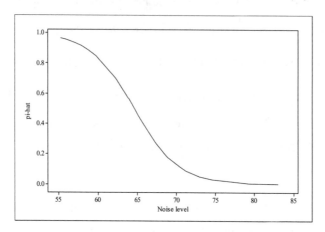

Section 13.3

27.

    a.   A scatter plot of the data indicated a quadratic regression model might be appropriate.

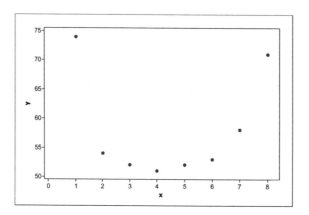

    b. $\hat{y} = 84.482 - 15.875(6) + 1.7679(6)^2 = 52.88$; residual $= y_6 - \hat{y}_6 = 53 - 52.88 = .12$

    c. $SST = \Sigma y_i^2 - \dfrac{(\Sigma y_i)^2}{n} = 586.88$, so $R^2 = 1 - \dfrac{61.77}{586.88} = .895$.

    d.   None of the standardized residuals exceeds 2 in magnitude, suggesting none of the observations are outliers. The ordered z percentiles needed for the normal probability plot are -1.53, $-.89$, $-.49$, $-.16$, .16, .49, .89, and 1.53. The normal probability plot below does not exhibit any troublesome features.

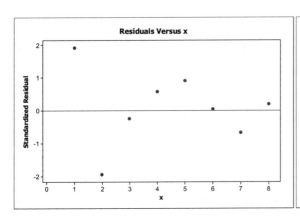
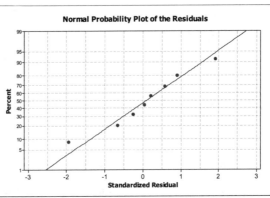

e. $\hat{\mu}_{Y\cdot6} = 52.88$ (from **b**) and $t_{.025,n-3} = t_{.025,5} = 2.571$, so the CI is

$52.88 \pm (2.571)(1.69) = 52.88 \pm 4.34 = (48.54, 57.22)$.

f. $SSE = 61.77$, so $s^2 = \dfrac{61.77}{5} = 12.35$ and $s\{pred\}\ \sqrt{12.35 + (1.69)^2} = 3.90$. The PI is

$52.88 \pm (2.571)(3.90) = 52.88 \pm 10.03 = (42.85, 62.91)$.

29.

a. The table below displays the y–values, fits, and residuals. From this, $SSE = \sum e^2 = 16.8$, $s^2 = SSE/(n - 3) = 4.2$, and $s = 2.05$.

y	$\hat{y}$	$e = y - \hat{y}$
81	82.1342	−1.13420
83	80.7771	2.22292
79	79.8502	−0.85022
75	72.8583	2.14174
70	72.1567	−2.15670
43	43.6398	−0.63985
22	21.5837	0.41630

b. $SST = \sum (y - \bar{y})^2 = \sum (y - 64.71)^2 = 3233.4$, so $R^2 = 1 - SSE/SST = 1 - 16.8/3233.4 = .995$, or 99.5%. 995% of the variation in free–flow can be explained by the quadratic regression relationship with viscosity.

c. We want to test the hypotheses $H_0: \beta_2 = 0$ v. $H_a: \beta_2 \neq 0$. Assuming all inference assumptions are met, the relevant t statistic is $t = \dfrac{-.0031662 - 0}{.0004835} = -6.55$. At $n - 3 = 4$df, the corresponding P–value is $2P(|t| > 6.55) < .004$. At any reasonable significance level, we would reject H_0 and conclude that the quadratic predictor indeed belongs in the regression model.

d. Two intervals with at least 95% simultaneous confidence requires individual confidence equal to 100% − 5%/2 = 97.5%. To use the t–table, round up to 98%: $t_{.01,4} = 3.747$. The two confidence intervals are $2.1885 \pm 3.747(.4050) = (.671, 3.706)$ for β_1 and $-.0031662 \pm 3.747(.0004835) = (-.00498, -.00135)$ for β_2. [In fact, we are at least 9<u>6</u>% confident β_1 and β_2 lie in these intervals.]

e. Plug into the regression equation to get $\hat{y} = 72.858$. Then a 95% CI for $\mu_{Y \cdot 400}$ is $72.858 \pm 3.747(1.198)$ $= (69.531, 76.186)$. For the PI, $s\{pred\} = \sqrt{s^2 + s_{\hat{y}}^2} = \sqrt{4.2 + (1.198)^2} = 2.374$, so a 95% PI for Y when $x = 400$ is $72.858 \pm 3.747(2.374) = (66.271, 79.446)$.

31.

a. $R^2 = 98.0\%$ or $.980$. This means 98.0% of the observed variation in energy output can be attributed to the model relationship.

b. For a quadratic model, adjusted $R^2 = \dfrac{(n-1)R^2 - k}{n-1-k} = \dfrac{(24-1)(.780) - 2}{24-1-2} = .759$, or 75.9%. (A more precise answer, from software, is 75.95%.) The adjusted R^2 value for the <u>cubic</u> model is 97.7%, as seen in the output. This suggests that the cubic term greatly improves the model: the cost of adding an extra parameter is more than compensated for by the improved fit.

c. To test the utility of the cubic term, the hypotheses are $H_0: \beta_3 = 0$ versus $H_0: \beta_3 \neq 0$. From the Minitab output, the test statistic is $t = 14.18$ with a P-value of $.000$. We strongly reject H_0 and conclude that the cubic term is a statistically significant predictor of energy output, even in the presence of the lower terms.

d. Plug $x = 30$ into the cubic estimated model equation to get $\hat{y} = 6.44$. From software, a 95% CI for $\mu_{Y \cdot 30}$ is $(6.31, 6.57)$. Alternatively, $\hat{y} \pm t_{.025,20} s_{\hat{y}} = 6.44 \pm 2.086(.0611)$ also gives $(6.31, 6.57)$. Next, a 95% PI for $Y \cdot 30$ is $(6.06, 6.81)$ from software. Or, using the information provided, $\hat{y} \pm t_{.025,20} \sqrt{s^2 + s_{\hat{y}}^2} = 6.44 \pm$ $2.086 \sqrt{(.1684)^2 + (.0611)^2}$ also gives $(6.06, 6.81)$. The value of s comes from the Minitab output, where $s = .168354$.

e. The null hypothesis states that the true mean energy output when the temperature difference is 35°K is equal to 5W; the alternative hypothesis says this isn't true.
Plug $x = 35$ into the cubic regression equation to get $\hat{y} = 4.709$. Then the test statistic is
$t = \dfrac{4.709 - 5}{.0523} \approx -5.6$, and the two-tailed P-value at df $= 20$ is approximately $2(.000) = .000$. Hence, we strongly reject H_0 (in particular, $.000 < .05$) and conclude that $\mu_{Y \cdot 35} \neq 5$.

Alternatively, software or direct calculation provides a 95% CI for $\mu_{Y \cdot 35}$ of $(4.60, 4.82)$. Since this CI does not include 5, we can reject H_0 at the .05 level.

33.

a. $\bar{x} = 20$ and $s_x = 10.8012$ so $x' = \dfrac{x - 20}{10.8012}$. For $x = 20$, $x' = 0$, and $\hat{y} = \hat{\beta}_0^* = .9671$. For $x = 25$, $x' = .4629$, so $\hat{y} = .9671 - .0502(.4629) - .0176(.4629)^2 + .0062(.4629)^3 = .9407$.

b. $\hat{y} = .9671 - .0502\left(\dfrac{x - 20}{10.8012}\right) - .0176\left(\dfrac{x - 20}{10.8012}\right)^2 + .0062\left(\dfrac{x - 20}{10.8012}\right)^3$
$.00000492x^3 - .000446058x^2 + .007290688x + .96034944$.

c. $t = \dfrac{.0062}{.0031} = 2.00$. We reject H_0 if either $t \geq t_{.025, n-4} = t_{.025,3} = 3.182$ or if $t \leq -3.182$. Since

2.00 is neither ≥ 3.182 nor ≤ -3.182, we cannot reject H_0; the cubic term should be deleted.

d. $SSE = \Sigma\left(y_i - \hat{y}_i\right)$ and the $\hat{y}_i$'s are the same from the standardized as from the unstandardized model, so SSE, SST, and R^2 will be identical for the two models.

e. $\Sigma y_i^2 = 6.355538$, $\Sigma y_i = 6.664$, so SST = .011410. For the quadratic model $R^2 = .987$ and for the cubic model, $R^2 = .994$; The two R^2 values are very close, suggesting intuitively that the cubic term is relatively unimportant.

35. $Y' = \ln(Y) = \ln\alpha + \beta x + \gamma x^2 + \ln(\varepsilon) = \beta_0 + \beta_1 x + \beta_2 x^2 + \varepsilon'$ where $\varepsilon' = \ln(\varepsilon)$, $\beta_0 = \ln(\alpha)$, $\beta_1 = \beta$, and $\beta_2 = \gamma$. That is, we should fit a quadratic to $(x, \ln(y))$. The resulting estimated quadratic (from computer output) is $2.00397 + .1799x - .0022x^2$, so $\hat{\beta} = .1799$, $\hat{\gamma} = -.0022$, and $\hat{\alpha} = e^{2.0397} = 7.6883$. (The ln(y)'s are 3.6136, 4.2499, 4.6977, 5.1773, and 5.4189, and the summary quantities can then be computed as before.)

Section 13.4

37.

a. The mean value of y when $x_1 = 50$ and $x_2 = 3$ is $\mu_{y \cdot 50, 3} = -.800 + .060(50) + .900(3) = 4.9$ hours.

b. When the number of deliveries (x_2) is held fixed, then average change in travel time associated with a one–mile (i.e. one unit) increase in distance traveled (x_1) is .060 hours. Similarly, when distance traveled (x_1) is held fixed, then the average change in travel time associated with on extra delivery (i.e., a one unit increase in x_2) is .900 hours.

c. Under the assumption that y follows a normal distribution, the mean and standard deviation of this distribution are 4.9 (because $x_1 = 50$ and $x_2 = 3$) and $\sigma = .5$ (since the standard deviation is assumed to be constant regardless of the values of x_1 and x_2). Therefore

$$P(y \leq 6) = P\left(z \leq \frac{6 - 4.9}{.5}\right) = P(z \leq 2.20) = .9861.$$ That is, in the long run, about 98.6% of all

days will result in a travel time of at most 6 hours.

39.

a. For $x_1 = 2$, $x_2 = 8$ (remember the units of x_2 are in 1000,s) and $x_3 = 1$ (since the outlet has a drive–up window) the average sales are $\hat{y} = 10.00 - 1.2(2) + 6.8(8) + 15.3(1) = 77.3$ (i.e., $77,300).

b. For $x_1 = 3$, $x_2 = 5$, and $x_3 = 0$ the average sales are $\hat{y} = 10.00 - 1.2(3) + 6.8(5) + 15.3(0) = 40.4$ (i.e., $40,400).

c. When the number of competing outlets (x_1) and the number of people within a 1–mile radius (x_2) remain fixed, the expected sales will increase by $15,300 when an outlet has a drive–up window.

41. $H_0 : \beta_1 = \beta_2 = ... = \beta_6 = 0$ vs. H_a: at least one among $\beta_1, ..., \beta_6$ is not zero. The test statistic is $F =$

$\dfrac{R^2 / k}{(1 - R^2) / (n - k - 1)}$. H_0 will be rejected if $f \geq F_{.05,6,30} = 2.42$. $f = \dfrac{.83 / 6}{(1 - .83) / 30} = 24.41$. Because $24.41 \geq$ 2.42, H_0 is rejected and the model is judged useful.

43.

   a. $x_1 = 2.6$, $x_2 = 250$, and $x_1 x_2 = (2.6)(250) = 650$, so
      $\hat{y} = 185.49 - 45.97(2.6) - 0.3015(250) + 0.0888(650) = 48.313$

   b. No, it is not legitimate to interpret β_1 in this way. It is not possible to increase by 1 unit the cobalt content, x_1, while keeping the interaction predictor, x_3, fixed. When x_1 changes, so does x_3, since $x_3 = x_1 x_2$.

   c. Yes, there appears to be a useful linear relationship between y and the predictors. We determine this by observing that the P-value corresponding to the model utility test is $< .0001$ (F test statistic = 18.924).

   d. We wish to test $H_0 : \beta_3 = 0$ vs. $H_a : \beta_3 \neq 0$. The test statistic is $t = 3.496$, with a corresponding P-value of .0030. Since the P-value is $< \alpha = .01$, we reject H_0 and conclude that the interaction predictor does provide useful information about y.

   e. A 95% C.I. for the mean value of surface area under the stated circumstances requires the following quantities: $\hat{y} = 185.49 - 45.97(2) - 0.3015(500) + 0.0888(2)(500) = 31.598$. Next, $t_{.025,16} = 2.120$, so the 95% confidence interval is $31.598 \pm (2.120)(4.69) = 31.598 \pm 9.9428 = (21.6552, 41.5408)$.

45.

   a. The hypotheses are $H_0: \beta_1 = \beta_2 = \beta_3 = \beta_4 = 0$ vs. H_a: at least one $\beta_i \neq 0$. The test statistic is $f =$
      $\dfrac{R^2 / k}{(1 - R^2) / (n - k - 1)} = \dfrac{.946 / 4}{(1 - .946) / 20} = 87.6 \geq F_{.001,4,20} = 7.10$ (the smallest available significance level from Table A.9), so we can reject H_0 at any significance level. We conclude that at least one of the four predictor variables appears to provide useful information about tenacity.

   b. The adjusted R^2 value is $1 - \dfrac{n-1}{n - (k+1)}\left(\dfrac{SSE}{SST}\right) = 1 - \dfrac{n-1}{n - (k+1)}(1 - R^2) = 1 - \dfrac{24}{20}(1 - .946) = .935$, which does not differ much from $R^2 = .946$.

   c. The estimated average tenacity when $x_1 = 16.5$, $x_2 = 50$, $x_3 = 3$, and $x_4 = 5$ is
      $\hat{y} = 6.121 - .082(16.5) + .113(50) + .256(3) - .219(5) = 10.091$. For a 99% C.I., $t_{.005,20} = 2.845$, so the interval is $10.091 \pm 2.845(.350) = (9.095, 11.087)$. Therefore, when the four predictors are as specified in this problem, the true average tenacity is estimated to be between 9.095 and 11.087.

47.

a. For a 1% increase in the percentage plastics, we would expect a 28.9 kcal/kg increase in energy content. Also, for a 1% increase in the moisture, we would expect a 37.4 kcal/kg decrease in energy content. Both of these assume we have accounted for the linear effects of the other three variables.

b. The appropriate hypotheses are $H_0 : \beta_1 = \beta_2 = \beta_3 = \beta_4 = 0$ vs. H_a : at least one $\beta_i \neq 0$. The value of the F test statistic is 167.71, with a corresponding p–value that is extremely small. So, we reject H_0 and conclude that at least one of the four predictors is useful in predicting energy content, using a linear model.

c. $H_0 : \beta_3 = 0$ vs. $H_a : \beta_3 \neq 0$. The value of the t test statistic is t = 2.24, with a corresponding p–value of .034, which is less than the significance level of .05. So we can reject H_0 and conclude that percentage garbage provides useful information about energy consumption, given that the other three predictors remain in the model.

d. $\hat{y} = 2244.9 + 28.925(20) + 7.644(25) + 4.297(40) - 37.354(45) = 1505.5$, and $t_{.025,25} = 2.060$. So a 95% C.I for the true average energy content under these circumstances is $1505.5 \pm (2.060)(12.47) = 1505.5 \pm 25.69 = (1479.8, 1531.1)$. Because the interval is reasonably narrow, we would conclude that the mean energy content has been precisely estimated.

e. A 95% prediction interval for the energy content of a waste sample having the specified characteristics is $1505.5 \pm (2.060)\sqrt{(31.48)^2 + (12.47)^2} = 1505.5 \pm 69.75 = (1435.7, 1575.2)$.

49.

a. $\hat{\mu}_{y \cdot 18,9,43} = 21.967 = 96.8303$, and residual = 91 – 96.8303 = –5.8303.

b. $H_0: \beta_1 = \beta_2 = 0$ vs. H_a: at least one $\beta_i \neq 0$; $f = \dfrac{R^2 / k}{(1 - R^2)/(n - k - 1)} = \dfrac{.798/2}{(1 - .798)/9} = 14.90 \geq F_{.05,2,9} = 8.02$, so we reject H_0. The model appears useful.

c. $96.8303 \pm (2.262)(8.20) = (78.28, 115.38)$

d. $96.8303 \pm (2.262)\sqrt{24.45^2 + 8.20^2} = (38.50, 155.16)$

e. We find the center of the given 95% interval, 93.875, and half of the width, 57.845. This latter value is equal to $t_{.025,9}(s_{\hat{y}}) = 2.262(s_{\hat{y}})$, so $s_{\hat{y}} = 25.5725$. Then the 90% interval is $93.785 \pm (1.833)(25.5725) = (46.911, 140.659)$

f. With the P-value for $H_a: \beta_1 \neq 0$ being 0.208 (from given output), we would fail to reject H_0. This factor is not significant given x_2 is in the model.

g. With $R_k^2 = .768$ (full model) and $R_l^2 = .721$ (reduced model), we can use an alternative F statistic: $F = \dfrac{(R_k^2 - R_l^2)/(k - l)}{(1 - R_k^2)/(n - k - 1)}$. With n = 12, k = 2 and l = 1, we have $f = \dfrac{(.768 - .721)/(2 - 1)}{(1 - .768)/12 - 2 - 1} = 1.83$. $t^2 = (-1.36)^2 = 1.85$. The discrepancy can be attributed to rounding error.

51.

a. No, there is no pattern in the plots which would indicate that a transformation or the inclusion of other terms in the model would produce a substantially better fit.

b. $k = 5$, $n - (k+1) = 8$, so $H_0 : \beta_1 = ... = \beta_5 = 0$ will be rejected if $f \geq F_{.05,5,8} = 3.69$;

$$f = \frac{(.759)/5}{(.241)/8} = 5.04 \geq 3.69 \text{, so we reject } H_0.$$ At least one of the coefficients is not equal to zero.

c. When $x_1 = 8.0$ and $x_2 = 33.1$ the residual is $e = 2.71$ and the standardized residual is $e^* = .44$; since $e^* = e/(\text{sd of the residual})$, sd of residual $= e/e^* = 6.16$. Thus the estimated variance of $\hat{Y}$ is $(6.99)^2 - (6.16)^2 = 10.915$, so the estimated sd is 3.304. Since $\hat{y} = 24.29$ and $t_{.025,8} = 2.306$, the desired C.I. is $24.29 \pm 2.306(3.304) = (16.67, 31.91)$.

d. $F_{.05,3,8} = 4.07$, so $H_0 : \beta_3 = \beta_4 = \beta_5 = 0$ will be rejected if $f \geq 4.07$. With

$$SSE_k = (14-6)s^2 = 390.88, \quad f = \frac{(894.95 - 390.88)/3}{390.88/8} = 3.44.$$ Since $3.44 < 4.07$, H_0 cannot be rejected and the quadratic terms should all be deleted. (N.B.: this is not a modification which would be suggested by a residual plot.)

53. Some possible questions might be:
(1) Is this model useful in predicting deposition of poly-aromatic hydrocarbons? A test of model utility gives us an F = 84.39, with a P-value of 0.000. Thus, the model is useful.
(2) Is x_1 a significant predictor of y in the presence of x_2? A test of $H_0: \beta_1 = 0$ v. $H_a: \beta_1 \neq 0$ gives us a t = 6.98 with a P-value of 0.000., so this predictor is significant.
(3) A similar question, and solution for testing x_2 as a predictor yields a similar conclusion: With a P-value of 0.046, we would accept this predictor as significant if our significance level were anything larger than 0.046.

Section 13.5

55.

a. $\ln(Q) = Y = \ln(\alpha) + \beta \ln(a) + \gamma \ln(b) + \ln(\varepsilon) = \beta_0 + \beta_1 x_1 + \beta_2 x_2 + \varepsilon'$ where $x_1 = \ln(a), x_2 = \ln(b), \beta_0 = \ln(\alpha), \beta_1 = \beta, \beta_2 = \gamma$ and $\varepsilon' = \ln(\varepsilon)$. Thus we transform to $(y, x_1, x_2) = (\ln(Q), \ln(a), \ln(b))$ (take the natural log of the values of each variable) and do a multiple linear regression. A computer analysis gave $\hat{\beta}_0 = 1.5652$, $\hat{\beta}_1 = .9450$, and $\hat{\beta}_2 = .1815$. For a = 10 and b = .01, $x_1 = \ln(10) = 2.3026$ and $x_2 = \ln(.01) = -4.6052$, from which $\hat{y} = 2.9053$ and $\hat{Q} = e^{2.9053} = 18.27$.

b. Again taking the natural log, $Y = \ln(Q) = \ln(\alpha) + \beta a + \gamma b + \ln(\varepsilon)$, so to fit this model it is necessary to take the natural log of each Q value (and not transform a or b) before using multiple regression analysis.

c. We simply exponentiate each endpoint: $\left(e^{.217}, e^{1.755} \right) = (1.24, 5.78)$.

57.

k	R^2	Adj. R^2	$C_k = \dfrac{SSE_k}{s^2} + 2(k+1) - n$
1	.676	.647	138.2
2	.979	.975	2.7
3	.9819	.976	3.2
4	.9824		4

where $s^2 = 5.9825$

a. Clearly the model with $k = 2$ is recommended on all counts.

b. No. Forward selection would let x_4 enter first and would not delete it at the next stage.

59. The choice of a "best" model seems reasonably clear–cut. The model with 4 variables including all but the summerwood fiber variable would seem best. R^2 is as large as any of the models, including the 5 variable model. R^2 adjusted is at its maximum and CP is at its minimum. As a second choice, one might consider the model with $k = 3$ which excludes the summerwood fiber and springwood % variables.

61. If multicollinearity were present, at least one of the four R^2 values would be very close to 1, which is not the case. Therefore, we conclude that multicollinearity is not a problem in this data.

63. Before removing any observations, we should investigate their source (e.g., were measurements on that observation misread?) <u>and</u> their impact on the regression. To begin, Observation #7 deviates significantly from the pattern of the rest of the data (standardized residual = –2.62); if there's concern the PAH deposition was not measured properly, we might consider removing that point to improve the overall fit. If the observation was <u>not</u> mis–recorded, we should <u>not</u> remove the point.

We should also investigate Observation #6: Minitab gives $h_{66} = .846 > 3(2+1)/17$, indicating this observation has very high leverage. However, the standardized residual for #6 is not large, suggesting that it follows the regression pattern specified by the other observations. Its "influence" only comes from having a comparatively large x_1 value.

Supplementary Exercises

65.

    a.

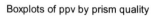

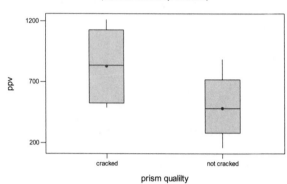

Boxplots of ppv by prism quality

(means are indicated by solid circles)

prism quality

A two–sample t confidence interval, generated by Minitab:

```
Two sample T for ppv

prism qu     N     Mean    StDev   SE Mean
cracked     12      827      295        85
not cracke  18      483      234        55

95% CI for mu (cracked   ) - mu (not cracke): ( 132,   557)
```

    b. The simple linear regression results in a significant model, r^2 is .577, but we have an extreme observation, with std resid = –4.11. Minitab output is below. Also run, but not included here was a model with an indicator for cracked/ not cracked, and for a model with the indicator and an interaction term. Neither improved the fit significantly.

```
The regression equation is
ratio = 1.00 -0.000018 ppv

Predictor        Coef       StDev           T         P
Constant      1.00161     0.00204      491.18     0.000
ppv        -0.00001827  0.00000295       -6.19     0.000

S = 0.004892    R-Sq = 57.7%    R-Sq(adj) = 56.2%

Analysis of Variance

Source          DF          SS          MS         F         P
Regression       1  0.00091571  0.00091571     38.26     0.000
Residual Error  28  0.00067016  0.00002393
Total           29  0.00158587

Unusual Observations
Obs      ppv     ratio         Fit   StDev Fit     Residual     St Resid
 29     1144  0.962000    0.980704    0.001786    -0.018704        -4.11R

R denotes an observation with a large standardized residual
```

67.

a. After accounting for all the other variables in the regression, we would expect the VO_2max to decrease by .0996, on average for each one–minute increase in the 1–mile walk time.

b. After accounting for all the other variables in the regression, we expect males to have a VO_2max that is .6566 L/min higher than females, on average.

c. $\hat{y} = 3.5959 + .6566(1) + .0096(170) - .0996(11) - .0880(140) = 3.67$. The residual is $\hat{y} = (3.15 - 3.67) = -.52$.

d. $R^2 = 1 - \dfrac{SSE}{SST} = 1 - \dfrac{30.1033}{102.3922} = .706$, or 70.6% of the observed variation in VO_2max can be attributed to the model relationship.

e. $H_0 : \beta_1 = \beta_2 = \beta_3 = \beta_4 = 0$ will be rejected in favor of H_a : at least one among $\beta_1, ..., \beta_4 \neq 0$, if $f \geq F_{.05,4,15} = 8.25$. With $f = \dfrac{(.706)/4}{(1-.706)/15} = 9.005 \geq 8.25$, so H_0 is rejected. It appears that the model specifies a useful relationship between VO_2max and at least one of the other predictors.

69.

a. Based on a scatter plot (below), a simple linear regression model would not be appropriate. Because of the slight, but obvious curvature, a quadratic model would probably be more appropriate.

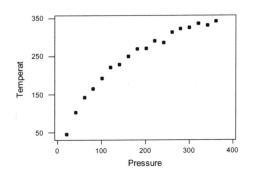

b. Using a quadratic model, a Minitab generated regression equation is $\hat{y} = 35.423 + 1.7191x - .0024753x^2$, and a point estimate of temperature when pressure is 200 is $\hat{y} = 280.23$. Minitab will also generate a 95% prediction interval of (256.25, 304.22). That is, we are confident that when pressure is 200 psi, a single value of temperature will be between 256.25 and 304.22°F.

71.

a. Using Minitab to generate the first order regression model, we test the model utility (to see if any of the predictors are useful), and with $f = 21.03$ and a p–value of .000, we determine that at least one of the predictors is useful in predicting palladium content. Looking at the individual predictors, the p–value associated with the pH predictor has value .169, which would indicate that this predictor is unimportant in the presence of the others.

b. Testing $H_0 : \beta_1 = ... = \beta_{20} = 0$ vs. H_a : at least one of the $\beta_i's \neq 0$. With calculated statistic $f = 6.29$, and p–value .002, this model is also useful at any reasonable significance level.

c. Testing $H_0 : \beta_6 = ... = \beta_{20} = 0$ vs. H_a: at least one of the listed $\beta_i's \neq 0$, the test statistic is

$f = \dfrac{(716.10 - 290.27)/(20 - 5)}{290.27(32 - 20 - 1)} = 1.07$. Using significance level .05, the rejection region would be

$f \geq F_{.05,15,11} = 2.72$. Since $1.07 < 2.72$, we fail to reject H_0 and conclude that all the quadratic and interaction terms should not be included in the model. They do not add enough information to make this model significantly better than the simple first order model.

d. Partial output from Minitab follows, which shows all predictors as significant at level .05:

```
The regression equation is
pdconc = - 305 + 0.405 niconc + 69.3 pH - 0.161 temp + 0.993 currdens
             + 0.355 pallcont - 4.14 pHsq
```

Predictor	Coef	StDev	T	P
Constant	-304.85	93.98	-3.24	0.003
niconc	0.40484	0.09432	4.29	0.000
pH	69.27	21.96	3.15	0.004
temp	-0.16134	0.07055	-2.29	0.031
currdens	0.9929	0.3570	2.78	0.010
pallcont	0.35460	0.03381	10.49	0.000
pHsq	-4.138	1.293	-3.20	0.004

73.

a. We wish to test $H_0 : \beta_1 = \beta_2 = 0$ vs. H_a: either β_1 or $\beta_2 \neq 0$. The test statistic is $f =$

$\dfrac{R^2/k}{(1-R^2)/(n-k-1)}$, where $k = 2$ for the quadratic model. The rejection region is

$f \geq F_{\alpha,k,n-k-1} = F_{.01,2,5} = 13.27$. $R^2 = 1 - \dfrac{.29}{202.88} = .9986$, giving $f = 1783 \gg 13.27$. The quadratic model is clearly useful.

b. The relevant hypotheses are $H_0 : \beta_2 = 0$ v. H_a: $\beta_2 \neq 0$. H_0 will be rejected at level .001 if either

$t \geq 6.869$ or $t \leq -6.869$ (df $= n - 3 = 5$). The test statistic value is $t = \dfrac{\hat{\beta}_2}{s_{\hat{\beta}_2}} =$

$\dfrac{-.00163141}{.00003391} = -48.1 \leq -6.869$, H_0 is rejected. The quadratic predictor should be retained.

c. No. R^2 is extremely high for the quadratic model, so the marginal benefit of including the cubic predictor would be essentially nil – and a scatter plot doesn't show the type of curvature associated with a cubic model.

d. $t_{.025,5} = 2.571$, and $\hat{\beta}_0 + \hat{\beta}_1(100) + \hat{\beta}_2(100)^2 = 21.36$, so the CI is $21.36 \pm 2.571(.1141) = 21.36 \pm .29$
$= (21.07, 21.65)$.

e. First, we need to figure out s^2 based on the information we have been given.
$s^2 = MSE = \frac{SSE}{df} = \frac{.29}{5} = .058$. Then, the 95% P.I. is
$21.36 \pm 2.571\sqrt{.058 + (.1141)^2} = 21.36 \pm 0.685 = (20.675, 22.045)$.

75.

a. $H_0 : \beta_1 = \beta_2 = 0$ will be rejected in favor of H_a : either β_1 or $\beta_2 \neq 0$ if $f = \dfrac{R^2/k}{(1-R^2)/(n-k-1)}$

$\geq F_{\alpha,k,n-k-1} = F_{.01,2,7} = 9.55$. $SST = \Sigma y^2 - \dfrac{(\Sigma y)}{n} = 264.5$, so $R^2 = 1 - \dfrac{26.98}{264.5} = .898$, and

$f = \dfrac{.898/2}{(.102)/7} = 30.8$. Because $30.8 \geq 9.55$ H_0 is rejected at significance level .01 and the quadratic

model is judged useful.

b. The hypotheses are $H_0 : \beta_2 = 0$ vs. $H_a : \beta_2 \neq 0$. The test statistic value is

$t = \dfrac{\hat{\beta}_2}{s_{\hat{\beta}_2}} = \dfrac{-2.3621}{.3073} = -7.69$, and $t_{.0005,7} = 5.408$, so H_0 is rejected at level .001. The quadratic

predictor should not be eliminated.

c. x = 1 here, and $\hat{\mu}_{Y\cdot1} = \hat{\beta}_0 + \hat{\beta}_1(1) + \hat{\beta}_2(1)^2 = 45.96$. $t_{.025,7} = 1.895$, giving the C.I.
$45.96 \pm (1.895)(1.031) = (44.01, 47.91)$.

77.

a. The hypotheses are $H_0: \beta_1 = \beta_2 = \beta_3 = \beta_4 = 0$ versus H_a: at least one $\beta_i \neq 0$. From the output, the F-statistic is $f = 4.06$ with a P-value of .029. Thus, at the .05 level we reject H_0 and conclude that at least one of the explanatory variables is a significant predictor of power.

b. Yes, a model with $R^2 = .834$ would appear to be useful. A formal model utility test can be performed:
$f = \dfrac{R^2/k}{(1-R^2)/[n-(k+1)]} = \dfrac{.834/3}{(1-.834)/[16-4]} = 20.1$, which is much greater than $F_{.05,3,12} = 3.49$. Thus,
the mode including $\{x_3, x_4, x_3x_4\}$ is useful.

We <u>cannot</u> use an F test to compare this model with the first-order model in (a), because neither model is a "subset" of the other. Compare $\{x_1, x_2, x_3, x_4\}$ to $\{x_3, x_4, x_3x_4\}$.

c. The hypotheses are $H_0: \beta_5 = \ldots = \beta_{10} = 0$ versus H_a: at least one of these $\beta_i \neq 0$, where β_5 through β_{10} are the coefficients for the six interaction terms. The "partial F test" statistic is
$f = \dfrac{(SSE_l - SSE_k)/(k-l)}{SSE_k/[n-(k+1)]} = \dfrac{(R_k^2 - R_l^2)/(k-l)}{(1-R_k^2)/[n-(k+1)]} = \dfrac{(.960-.596)/(10-4)}{(1-.960)/[16-(10+1)]} = 7.58$, which is greater

than $F_{.05,6,5} = 4.95$. Hence, we reject H_0 at the .05 level and conclude that at least one of the interaction terms is a statistically significant predictor of power, in the presence of the first-order terms.

79. There are obviously several reasonable choices in each case. In **a**, the model with 6 carriers is a defensible choice on all three grounds, as are those with 7 and 8 carriers. The models with 7, 8, or 9 carriers in **b** merit serious consideration. These models merit consideration because R_k^2, MSE_k, and C_k meet the variable selection criteria given in Section 13.5.

81.

   a. The relevant hypotheses are $H_0 : \beta_1 = ... = \beta_5 = 0$ vs. H$_a$: at least one among $\beta_1,...,\beta_5 \neq 0$. $f =$ $\dfrac{.827/5}{.173/11} = 106.1 \geq F_{.05,5,111} \approx 2.29$. Hence, H$_0$ is rejected in favor of the conclusion that there is a useful linear relationship between Y and at least one of the predictors.

   b. $t_{.05,111} = 1.66$, so the CI is $.041 \pm (1.66)(.016) = .041 \pm .027 = (.014, .068)$. β_1 is the expected change in mortality rate associated with a one-unit increase in the particle reading when the other four predictors are held fixed; we can be 90% confident that $.014 < \beta_1 < .068$.

   c. $H_0 : \beta_4 = 0$ will be rejected in favor of $H_a : \beta_4 \neq 0$ if $t = \dfrac{\hat{\beta}_4}{s_{\hat{\beta}_4}}$ is either ≥ 2.62 or ≤ -2.62.

      $t = \dfrac{.047}{.007} = 5.9 \geq 2.62$, so H$_0$ is rejected and this predictor is judged important.

   d. $\hat{y} = 19.607 + .041(166) + .071(60) + .001(788) + .041(68) + .687(.95) = 99.514$, and the corresponding residual is $103 - 99.514 = 3.486$.

CHAPTER 14

Section 14.1

1.

    a. We reject H_0 if the calculated χ^2 value is greater than or equal to the tabled value of $\chi^2_{\alpha,k-1}$ from Table A.7. Since $12.25 \geq \chi^2_{.05,4} = 9.488$, we would reject H_0.

    b. Since 8.54 is not $\geq \chi^2_{.01,3} = 11.344$, we would fail to reject H_0.

    c. Since 4.36 is not $\geq \chi^2_{.10,2} = 4.605$, we would fail to reject H_0.

    d. Since 10.20 is not $\geq \chi^2_{.01,5} = 15.085$, we would fail to reject H_0.

3.   Let p_1, p_2, p_3, p_4 denote the true proportion of all African American, Asian, Caucasian, and Hispanic characters in commercials (broadcast in the Philadelphia area), respectively. The null hypothesis is H_0: $p_1 = .177, p_2 = .032, p_3 = .734, p_4 = .057$ (that is, the proportions in commercials match census proportions). The alternative hypothesis is that at least one of these proportions is incorrect.

    The sample size is $n = 404$, so the expected counts under H_0 are $404(.177) = 71.508$, $404(.032) = 12.928$, $404(.734) = 296.536$, and $404(.057) = 23.028$. The resulting chi-squared goodness-of-fit statistic is

$$\chi^2 = \frac{(57 - 71.508)^2}{71.508} + \cdots + \frac{(6 - 23.028)^2}{23.028} = 19.6.$$

At df $= 4 - 1 = 3$, the P-value is less than .001 (since $19.6 > 16.26$). Hence, we strongly reject H_0 and conclude that at least one of the racial proportions in commercials is <u>not</u> a match to the census proportions.

5.   We will reject H_0 if the p-value $< .10$. The observed values, expected values, and corresponding χ^2 terms are :

Obs	4	15	23	25	38	21	32	14	10	8
Exp	6.67	13.33	20	26.67	33.33	33.33	26.67	20	13.33	6.67
χ^2	1.069	.209	.450	.105	.654	.163	1.065	1.800	.832	.265

$\chi^2 = 1.069 + \ldots + .265 = 6.612$. With df. $= 10 - 1 = 9$, our χ^2 value of 6.612 is less than $\chi^2_{.10,9} = 14.684$, so the p-value $> .10$ and we cannot reject H_0. There is no significant evidence that the data is not consistent with the previously determined proportions.

7. We test $H_o: p_1 = p_2 = p_3 = p_4 = .25$ vs. H_a: at least one proportion $\neq .25$, and df $= 3$. We will reject H_0 if the p-value $< .01$.

Cell	1	2	3	4
Observed	328	334	372	327
Expected	340.25	340.25	340.25	34.025
χ^2 term	.4410	.1148	2.9627	.5160

$\chi^2 = 4.0345$, and with 3 df., p-value $> .10$, so we fail to reject H_0. The data fails to indicate a seasonal relationship with incidence of violent crime.

9.

a. Denoting the 5 intervals by $[0, c_1), [c_1, c_2), \ldots, [c_4, \infty)$, we wish c_1 for which

$.2 = P(0 \le X \le c_1) = \int_0^{c_1} e^{-x} dx = 1 - e^{-c_1}$, so $c_1 = -\ln(.8) = .2231$. Then

$.2 = P(c_1 \le X \le c_2) \Rightarrow .4 = P(0 \le X_1 \le c_2) = 1 - e^{-c_2}$, so $c_2 = -\ln(.6) = .5108$. Similarly, $c_3 = -\ln(.4) = .0163$ and $c_4 = -\ln(.2) = 1.6094$. the resulting intervals are $[0, .2231), [.2231, .5108), [.5108, .9163), [.9163, 1.6094),$ and $[1.6094, \infty)$.

b. Each expected cell count is $40(.2) = 8$, and the observed cell counts are 6, 8, 10, 7, and 9, so

$\chi^2 = \left[\frac{(6-8)^2}{8} + \ldots + \frac{(9-8)^2}{8} \right] = 1.25$. Because 1.25 is not $\ge \chi^2_{.10,4} = 7.779$, even at level

.10 H_0 cannot be rejected; the data is quite consistent with the specified exponential distribution.

11.

a. The six intervals must be symmetric about 0, so denote the 4th, 5th and 6th intervals by $[0, a), [a, b), [b, \infty)$. a must be such that $\Phi(a) = .6667(\frac{1}{2} + \frac{1}{6})$, which from Table A.3 gives $a \approx .43$. Similarly, $\Phi(b) = .8333$ implies $b \approx .97$, so the six intervals are $(-\infty, -.97), [-.97, -.43), [-.43, 0), [0, .43), [.43, .97),$ and $[.97, \infty)$.

b. The six intervals are symmetric about the mean of .5. From **a**, the fourth interval should extend from the mean to .43 standard deviations above the mean, i.e., from .5 to $.5 + .43(.002)$, which gives $[.5, .50086)$. Thus the third interval is $[.5 - .00086, .5) = [.49914, .5)$. Similarly, the upper endpoint of the fifth interval is $.5 + .97(.002) = .50194$, and the lower endpoint of the second interval is $.5 - .00194 = .49806$. The resulting intervals are $(-\infty, .49806), [.49806, .49914), [.49914, .5), [.5, .50086), [.50086, .50194),$ and $[.50194, \infty)$.

c. Each expected count is $45(\frac{1}{6}) = 7.5$, and the observed counts are 13, 6, 6, 8, 7, and 5, so $\chi^2 = 5.53$. With 5 df., the p-value $> .10$, so we would fail to reject H_0 at any of the usual levels of significance. There is no significant evidence to suggest that the bolt diameters are not normally distributed with $\mu = .5$ and $\sigma = .002$.

Section 14.2

13. According to the stated model, the three cell probabilities are $(1 - p)^2$, $2p(1 - p)$, and p^2, so we wish the value of p which maximizes $(1 - p)^{2n_1} [2p(1 - p)]^{n_2} p^{2n_3}$. Proceeding as in example 14.6 gives

$\hat{p} = \dfrac{n_2 + 2n_3}{2n} = \dfrac{234}{2776} = .0843$. The estimated expected cell counts are then $n(1 - \hat{p})^2 = 1163.85$,

$n[2\hat{p}(1 - \hat{p})]^2 = 214.29$, $n\hat{p}^2 = 9.86$. This gives

$\chi^2 = \left[\dfrac{(1212 - 1163.85)^2}{1163.85} + \dfrac{(118 - 214.29)^2}{214.29} + \dfrac{(58 - 9.86)^2}{9.86} \right] = 280.3$. According to (14.15), H_0 will be

rejected if $\chi^2 \geq \chi^2_{\alpha, 2}$, and since $\chi^2_{.01,2} = 9.210$, H_0 is soundly rejected; the stated model is strongly contradicted by the data.

15. The part of the likelihood involving θ is $\left[(1 - \theta)^4 \right]^{n_1} \cdot \left[\theta(1 - \theta)^3 \right]^{n_2} \cdot \left[\theta^2 (1 - \theta)^2 \right]^{n_3} \cdot$

$\left[\theta^3 (1 - \theta) \right]^{n_4} \cdot \left[\theta^4 \right]^{n_5} = \theta^{n_2 + 2n_3 + 3n_4 + 4n_5} (1 - \theta)^{4n_1 + 3n_2 + 2n_3 + n_4} = \theta^{233} (1 - \theta)^{367}$, so the log-likelihood is

$233 \ln \theta + 367 \ln(1 - \theta)$. Differentiating and equating to 0 yields $\hat{\theta} = \dfrac{233}{600} = .3883$, and $(1 - \hat{\theta}) = .6117$

[note that the exponent on θ is simply the total # of successes (defectives here) in the $n = 4(150) = 600$ trials]. Substituting this $\hat{\theta}$ into the formula for p_i yields estimated cell probabilities .1400, .3555, .3385, .1433, and .0227. Multiplication by 150 yields the estimated expected cell counts are 21.00, 53.33, 50.78, 21.50, and 3.41. the last estimated expected cell count is less than 5, so we combine the last two categories into a single one (≥ 3 defectives), yielding estimated counts 21.00, 53.33, 50.78, 24.91, observed counts 26, 51, 47, 26, and $\chi^2 = 1.62$. With $df = 4 - 1 - 1 = 2$, since $1.62 < \chi^2_{.10,2} = 4.605$, the P-value $> .10$, and we do not reject H_0. The data suggests that the stated binomial distribution is plausible.

17. $\hat{\lambda} = \dfrac{380}{120} = 3.167$, so $\hat{p} = e^{-3.167} \dfrac{(3.167)^x}{x!}$.

x	0	1	2	3	4	5	6	≥ 7
$\hat{p}$	.0421	.1334	.2113	.2230	.1766	.1119	.0590	.0427
$n\hat{p}$	5.05	16.00	25.36	26.76	21.19	13.43	7.08	5.12
obs	24	16	16	18	15	9	6	16

The resulting value of $\chi^2 = 103.98$, and when compared to $\chi^2_{.01,7} = 18.474$, it is obvious that the Poisson model fits very poorly.

19. With $A = 2n_1 + n_4 + n_5$, $B = 2n_2 + n_4 + n_6$, and $C = 2n_3 + n_5 + n_6$, the likelihood is proportional to $\theta_1^A \theta_2^B (1 - \theta_1 - \theta_2)^C$, where $A + B + C = 2n$. Taking the natural log and equating both $\dfrac{\partial}{\partial \theta_1}$ and $\dfrac{\partial}{\partial \theta_2}$

to zero gives $\dfrac{A}{\theta_1} = \dfrac{C}{1 - \theta_1 - \theta_2}$ and $\dfrac{B}{\theta_2} = \dfrac{C}{1 - \theta_1 - \theta_2}$, whence $\theta_2 = \dfrac{B\theta_1}{A}$. Substituting this into the

first equation gives $\theta_1 = \dfrac{A}{A + B + C}$, and then $\theta_2 = \dfrac{B}{A + B + C}$. Thus $\hat{\theta}_1 = \dfrac{2n_1 + n_4 + n_5}{2n}$,

$\hat{\theta}_2 = \dfrac{2n_2 + n_4 + n_6}{2n}$, and $\left(1 - \hat{\theta}_1 - \hat{\theta}_2\right) = \dfrac{2n_3 + n_5 + n_6}{2n}$. Substituting the observed n_i's yields

$\hat{\theta}_1 = \dfrac{2(49) + 20 + 53}{400} = .4275$, $\hat{\theta}_2 = \dfrac{110}{400} = .2750$, and $\left(1 - \hat{\theta}_1 - \hat{\theta}_2\right) = .2975$, from which

$\hat{p}_1 = (.4275)^2 = .183$, $\hat{p}_2 = .076$, $\hat{p}_3 = .089$, $\hat{p}_4 = 2(.4275)(.275) = .235$, $\hat{p}_5 = .254$, $\hat{p}_6 = .164$.

Category	1	2	3	4	5	6
np	36.6	15.2	17.8	47.0	50.8	32.8
observed	49	26	14	20	53	38

This gives $\chi^2 = 29.1$. With $\chi^2_{.01,6-1-2} = \chi^2_{.01,3} = 11.344$, and $\chi^2_{.01,6-1} = \chi^2_{.01,5} = 15.085$, according to (14.15) H_0 must be rejected since $29.1 \geq 15.085$.

21. The Ryan-Joiner test p-value is larger than .10, so we conclude that the null hypothesis of normality cannot be rejected. This data could reasonably have come from a normal population. This means that it would be legitimate to use a one-sample t test to test hypotheses about the true average ratio.

23. Minitab gives r = .967, though the hand calculated value may be slightly different because when there are ties among the $x_{(i)}$'s, Minitab uses the same y_i for each $x_{(i)}$ in a group of tied values. $C_{10} = .9707$, and $c_{.05} = 9639$, so $.05 <$ p-value $< .10$. At the 5% significance level, one would have to consider population normality plausible.

Section 14.3

25. Let p_{ij} = the proportion of white clover in area of type i which has a type j mark (i = 1, 2; j = 1, 2, 3, 4, 5). The hypothesis H_0: $p_{1j} = p_{2j}$ for j = 1, ..., 5 will be rejected at level .01 if $\chi^2 \geq \chi^2_{.01,(2-1)(5-1)} = \chi^2_{.01,4} = 13.277$.

$\hat{E}_{ij}$	1	2	3	4	5		
1	449.66	7.32	17.58	8.79	242.65	726	$\chi^2 = 23.18$
2	471.34	7.68	18.42	9.21	254.35	761	
	921	15	36	18	497	1487	

Since $23.18 \geq 13.277$, H_0 is rejected.

204

27. With i = 1 identified with men and i = 2 identified with women, and j = 1, 2, 3 denoting the 3 categories L>R, L=R, L<R, we wish to test H_0: $p_{1j} = p_{2j}$ for j = 1, 2, 3 vs. H_a: p_{1j} not equal to p_{2j} for at least one j. The estimated cell counts for men are 17.95, 8.82, and 13.23 and for women are 39.05, 19.18, 28.77, resulting in $\chi^2 = 44.98$. With (2 – 1)(3 – 1) = 2 degrees of freedom, since $44.98 > \chi^2_{.005,2} = 10.597$, p-value < .005, which strongly suggests that H_0 should be rejected.

29.

a. The null hypothesis is H_0: $p_{1j} = p_{2j} = p_{3j}$ for j = 1, 2, 3, 4, where p_{ij} is the proportion of the ith population (natural scientists, social scientists, non-academics with graduate degrees) whose degree of spirituality falls into the jth category (very, moderate, slightly, not at all).

From the accompanying Minitab output, the test statistic value is $\chi^2 = 213.212$ with df = (3–1)(4–1) = 6, with an associated P-value of 0.000. Hence, we strongly reject H_0. These three populations are <u>not</u> homogeneous with respect to their degree of spirituality.

Chi-Square Test: Very, Moderate, Slightly, Not At All

```
Expected counts are printed below observed counts
Chi-Square contributions are printed below expected counts

          Very   Moderate   Slightly   Not At All   Total
   1        56        162        198          211      627
          78.60     195.25     183.16       170.00
          6.497      5.662      1.203        9.889

   2        56        223        243          239      761
          95.39     236.98     222.30       206.33
         16.269      0.824      1.928        5.173

   3       109        164         74           28      375
          47.01     116.78     109.54       101.67
         81.752     19.098     11.533       53.384

Total      221        549        515          478     1763

Chi-Sq = 213.212, DF = 6, P-Value = 0.000
```

b. We're now testing H_0: $p_{1j} = p_{2j}$ for j = 1, 2, 3, 4 under the same notation. The accompanying Minitab output shows $\chi^2 = 3.091$ with df = (2–1)(4–1) = 3 and an associated P-value of 0.378. Since this is larger than any reasonable significance level, we fail to reject H_0. The data provides no statistically significant evidence that the populations of social and natural scientists differ with respect to degree of spirituality.

Chi-Square Test: Very, Moderate, Slightly, Not At All

```
Expected counts are printed below observed counts
Chi-Square contributions are printed below expected counts

        Very  Moderate  Slightly  Not At All  Total
   1      56       162       198         211    627
        50.59    173.92    199.21      203.28
        0.578     0.816     0.007       0.293

   2      56       223       243         239    761
        61.41    211.08    241.79      246.72
        0.476     0.673     0.006       0.242

Total    112       385       441         450   1388

Chi-Sq = 3.091, DF = 3, P-Value = 0.378
```

31.

a. The accompanying table shows the proportions of male and female smokers in the sample who began smoking at the ages specified. (The male proportions were calculated by dividing the counts by the total of 96; for females, we divided by 93.) The patterns of the proportions seems to be different, suggesting there does exist an association between gender and age at first smoking.

		Gender	
		Male	**Female**
	<16	0.26	0.11
Age	**16–17**	0.25	0.34
	18–20	0.29	0.18
	>20	0.20	0.37

b. The hypotheses, in words, are H_0: gender and age at first smoking are independent, versus H_a: gender and age at first smoking are associated. The accompanying Minitab output provides a test statistic value of $\chi^2 = 14.462$ at df = (2–1)(4–1) = 3, with an associated P-value of 0.002. Hence, we would reject H_0 at both the .05 and .01 levels. We have evidence to suggest an association between gender and age at first smoking.

(Minitab output appears on the next page.)

Chi-Square Test: Male, Female

```
Expected counts are printed below observed counts
Chi-Square contributions are printed below expected counts

        Male   Female   Total
   1      25       10      35
        17.78    17.22
        2.934    3.029

   2      24       32      56
        28.44    27.56
        0.694    0.717

   3      28       17      45
        22.86    22.14
        1.157    1.194

   4      19       34      53
        26.92    26.08
        2.330    2.406

Total     96       93     189

Chi-Sq = 14.462, DF = 3, P-Value = 0.002
```

33.
$$\chi^2 = \Sigma\Sigma\frac{\left(N_{ij} - \hat{E}_{ij}\right)^2}{\hat{E}_{ij}} = \Sigma\Sigma\frac{N_{ij}^2 - 2\hat{E}_{ij}N_{ij} + \hat{E}_{ij}^2}{\hat{E}_{ij}} = \Sigma\Sigma\frac{N_{ij}^2}{\hat{E}_{ij}} - 2\Sigma\Sigma N_{ij} + \Sigma\Sigma\hat{E}_{ij}, \text{ but}$$

$\Sigma\Sigma\hat{E}_{ij} = \Sigma\Sigma N_{ij} = n$, so $\chi^2 = \Sigma\Sigma\dfrac{N_{ij}^2}{\hat{E}_{ij}} - n$. This formula is computationally efficient because there is

only one subtraction to be performed, which can be done as the last step in the calculation.

35.
With p_{ij} denoting the common value of p_{ij1}, p_{ij2}, p_{ij3}, p_{ij4} (under H_0), $\hat{p}_{ij} = \dfrac{n_{ij.}}{n}$ and $\hat{E}_{ijk} = \dfrac{n_k n_{ij.}}{n}$, where

$n_{ij.} = \displaystyle\sum_{k=1}^{4} n_{ijk}$ and $n = \displaystyle\sum_{k=1}^{4} n_k$. With four different tables (one for each region), there are $4(9-1) = 32$

freely determined cell counts. Under H_0, the nine parameters p_{11}, …, p_{33} must be estimated, but $\Sigma\Sigma p_{ij} = 1$,
so only 8 independent parameters are estimated, giving χ^2 df $= 32 - 8 = 24$. Note: this is really a test of
homogeneity for 4 strata, each with 3x3=9 categories. Hence, df $= (4-1)(9-1) = 24$.

Supplementary Exercises

37. There are 3 categories here – firstborn, middleborn, (2nd or 3rd born), and lastborn. With p_1, p_2, and p_3 denoting the category probabilities, we wish to test H_0: $p_1 = .25$, $p_2 = .50$ ($p_2 = P(2^{nd}$ or 3rd born) $= .25 + .25 = .50$), $p_3 = .25$. H_0 will be rejected at significance level .05 if $\chi^2 \geq \chi^2_{.05,2} = 5.992$. The expected counts are $(31)(.25) = 7.75$, $(31)(.50) = 15.5$, and 7.75, so $\chi^2 = \frac{(12-7.75)^2}{7.75} + \frac{(11-15.5)^2}{15.5} + \frac{(8-7.75)^2}{7.75} = 3.65$.

Because $3.65 < 5.992$, H_0 is not rejected. The hypothesis of equiprobable birth order appears quite plausible.

39. This is a test of homogeneity, with hypotheses H_0: proportions falling into these experience categories are the same for men and women, versus H_a: proportions falling into these experience categories are different for men and women. Df = 4, and we reject H_0 if $\chi^2 \geq \chi^2_{.01,4} = 13.277$.

Gender	Years of Experience				
	1 – 3	4 – 6	7 – 9	10 – 12	13 +
Male Observed	202	369	482	361	811
Expected	285.56	409.83	475.94	347.04	706.63
$(O-E)^2/E$	24.451	4.068	.077	.562	15.415
Female Observed	230	251	238	164	258
Expected	146.44	210.17	244.06	177.96	362.37
$(O-E)^2/E$	47.680	7.932	.151	1.095	30.061

$\chi^2 = \Sigma \frac{(O-E)^2}{E} = 131.492$. Reject H_0. The proportions of male and female coaches falling into these experience categories are very different. In particular, women have higher than expected counts in the beginning category (1 – 3 years) and lower than expected counts in the more experienced category (13+ years).

41. The null hypothesis H_0: $p_{ij} = p_{i.} \, p_{.j}$ states that level of parental use and level of student use are independent in the population of interest. The test is based on $(3-1)(3-1) = 4$ df.

Estimated expected counts

119.3	57.6	58.1	235
82.8	33.9	40.3	163
23.9	11.5	11.6	47
226	109	110	445

The calculated value of $\chi^2 = 22.4$. Since $22.4 > \chi^2_{.005,4} = 14.860$, p-value < .005, so H_0 should be rejected at any significance level greater than .005. Parental and student use level do not appear to be independent.

43. This is a test of homogeneity: H_0: $p_{1j} = p_{2j} = p_{3j}$ for $j = 1, 2, 3, 4, 5$. The given SPSS output reports the calculated $\chi^2 = 70.64156$ and accompanying p-value (significance) of .0000. We reject H_0 at any significance level. The data strongly supports that there are differences in perception of odors among the three areas.

45. $\left(n_1 - np_{10}\right)^2 = \left(np_{10} - n_1\right)^2 = \left(n - n_1 - n\left(1 - p_{10}\right)\right)^2 = \left(n_2 - np_{20}\right)^2$. Therefore

$$\chi^2 = \frac{\left(n_1 - np_{10}\right)^2}{np_{10}} + \frac{\left(n_2 - np_{20}\right)^2}{np_{20}} = \frac{\left(n_1 - np_{10}\right)^2}{n_2}\left(\frac{n}{p_{10}} + \frac{n}{p_{20}}\right)$$

$$= \left(\frac{n_1}{n} - p_{10}\right)^2 \cdot \left(\frac{n}{p_{10}p_{20}}\right) = \frac{\left(\hat{p}_1 - p_{10}\right)^2}{p_{10}p_{20}/n} = z^2.$$

47.

a. Our hypotheses are H_0: no difference in proportion of concussions among the three groups v. H_a: there is a difference in proportion of concussions among the three groups.

Observed	Concussion	No Concussion	Total
Soccer	45	46	91
Non Soccer	28	68	96
Control	8	45	53
Total	81	159	240

Expected	Concussion	No Concussion	Total
Soccer	30.7125	60.2875	91
Non Soccer	32.4	63.6	96
Control	17.8875	37.1125	53
Total	81	159	240

$$\chi^2 = \frac{\left(45 - 30.7125\right)^2}{30.7125} + \frac{\left(46 - 60.2875\right)^2}{60.2875} + \frac{\left(28 - 32.4\right)^2}{32.4} + \frac{\left(68 - 63.6\right)^2}{63.6}$$

$$+ \frac{\left(8 - 17.8875\right)^2}{17.8875} + \frac{\left(45 - 37.1125\right)^2}{37.1125} = 19.1842.$$ The df for this test is $(I - 1)(J - 1) = 2$, so we reject H_0 if $\chi^2 > \chi^2_{.05,2} = 5.99$. $19.1842 > 5.99$, so we reject H_0. There is a difference in the proportion of concussions based on whether a person plays soccer.

b. The sample correlation of $r = -.220$ indicates a weak negative association between "soccer exposure" and immediate memory recall. We can formally test the hypotheses H_0: $\rho = 0$ vs H_a: $\rho < 0$. The test statistic is $t = \frac{r\sqrt{n-2}}{\sqrt{1-r^2}} = \frac{-.22\sqrt{89}}{\sqrt{1-.22^2}} = -2.13$. At significance level $\alpha = .01$, we would fail to reject H_0 and conclude that there is no significant evidence of negative association in the population.

c. We will test to see if the average score on a controlled word association test is the same for soccer and non-soccer athletes. H_0: $\mu_1 = \mu_2$ vs H_a: $\mu_1 \neq \mu_2$. Since the two sample standard deviations are very close, we will use a pooled-variance two-sample t test. From Minitab, the test statistic is $t = -0.91$, with an associated P-value of 0.366 at 80 df. We clearly fail to reject H_0 and conclude that there is no statistically significant difference in the average score on the test for the two groups of athletes.

d. Our hypotheses for ANOVA are H_0: all means are equal vs H_a: not all means are equal. The test statistic is $f = \dfrac{MSTr}{MSE}$. $SSTr = 91(.30 - .35)^2 + 96(.49 - .35)^2 + 53(.19 - .35)^2 = 3.4659$

$MSTr = \dfrac{3.4659}{2} = 1.73295$ $SSE = 90(.67)^2 + 95(.87)^2 + 52(.48)^2 = 124.2873$ and

$MSE = \dfrac{124.2873}{237} = .5244$. Now, $f = \dfrac{1.73295}{.5244} = 3.30$. Using df = (2,200) from table A.9, the p value is between .01 and .05. At significance level .05, we reject the null hypothesis. There is sufficient evidence to conclude that there is a difference in the average number of prior non-soccer concussions between the three groups.

49. According to Benford's law, the probability a lead digit equals x is given by $\log_{10}(1 + 1/x)$ for $x = 1, \ldots, 9$. Let p_i = the proportion of Fibonacci numbers whose lead digit is i ($i = 1, \ldots, 9$). We wish to perform a goodness-of-fit test H_0: $p_i = \log_{10}(1 + 1/i)$ for $i = 1, \ldots, 9$. (The alternative hypothesis is that Benford's formula is incorrect for at least one category.) The table below summarizes the results of the test.

Digit	1	2	3	4	5	6	7	8	9
Obs. #	25	16	11	7	7	5	4	6	4
Exp. #	25.59	14.97	10.62	8.24	6.73	5.69	4.93	4.35	3.89

Expected counts are calculated by $np_i = 85 \log_{10}(1 + 1/i)$. Some of the expected counts are too small, so combine 6 and 7 into one category (obs = 9, exp = 10.62); do the same to 8 and 9 (obs = 10, exp = 8.24).

The resulting chi-squared statistic is $\chi^2 = \dfrac{(25 - 25.59)^2}{25.59} + \cdots + \dfrac{(10 - 8.24)^2}{8.24} = 0.92$ at df = 7 – 1 = 6 (since there are 7 categories after the earlier combining). Software provides a P-value of .988!

We certainly do not reject H_0 — the lead digits of the Fibonacci sequence are highly consistent with Benford's law.

CHAPTER 15

Section 15.1

1. We test $H_0 : \mu = 100$ vs. $H_a : \mu \neq 100$. The test statistic is s_+ = sum of the ranks associated with the positive values of $(x_i - 100)$, and we reject H_o at significance level .05 if $s_+ \geq 64$. (from Table A.13, n = 12, with $\alpha/2 = .026$, which is close to the desired value of .025), or if $s_+ \leq \dfrac{12(13)}{2} - 64 = 78 - 64 = 14$.

x_i	$(x_i - 100)$	ranks
105.6	5.6	7*
90.9	-9.1	12
91.2	-8.8	11
96.9	-3.1	3
96.5	-3.5	5
91.3	-8.7	10
100.1	0.1	1*
105	5	6*
99.6	-0.4	2
107.7	7.7	9*
103.3	3.3	4*
92.4	-7.6	8

    $S_+ = 27$, and since 27 is neither ≥ 64 nor ≤ 14, we do not reject H_o. There is not enough evidence to suggest that the mean is something other than 100.

3. We test $H_0 : \mu = 7.39$ vs. $H_a : \mu \neq 7.39$, so a two tailed test is appropriate. With n = 14 and $\alpha/2 = .025$, Table A.13 indicates that H_o should be rejected if either $s_+ \geq 84\, or \leq 21$. The $(x_i - 7.39)$'s are -.37, -.04, -.05, -.22, -.11, .38, -.30, -.17, .06, -.44, .01, -.29, -.07, and -.25, from which the ranks of the three positive differences are 1, 4, and 13. Since $s_+ = 18 \leq 21$, H_o is rejected at level .05.

5. The data is paired, and we wish to test $H_0 : \mu_D = 0$ vs. $H_a : \mu_D \neq 0$. With n = 12 and $\alpha = .05$, H_o should be rejected if either $s_+ \geq 64$ or if $s_+ \leq 14$.

d_i	-.3	2.8	3.9	.6	1.2	-1.1	2.9	1.8	.5	2.3	.9	2.5
rank	1	10*	12*	3*	6*	5	11*	7*	2*	8*	4*	9*

    $s_+ = 72$, and $72 \geq 64$, so H_o is rejected at level .05. In fact for $\alpha = .01$, the critical value is c = 71, so even at level .01 H_o would be rejected.

211

7. $H_0 : \mu_D = .20$ vs. $H_a : \mu_D > .20$, where $\mu_D = \mu_{outdoor} - \mu_{indoor}$. $\alpha = .05$, and because n = 33,

we can use the large sample test. The test statistic is $Z = \dfrac{s_+ - \frac{n(n+1)}{4}}{\sqrt{\frac{n(n+1)(2n+1)}{24}}}$, and we reject H_0 if $z \geq 1.96$.

d_i	$d_i - .2$	rank	d_i	$d_i - .2$	rank	d_i	$d_i - .2$	rank
0.22	0.02	2	0.15	-0.05	5.5	0.63	0.43	23
0.01	-0.19	17	1.37	1.17	32	0.23	0.03	4
0.38	0.18	16	0.48	0.28	21	0.96	0.76	31
0.42	0.22	19	0.11	-0.09	8	0.2	0	1
0.85	0.65	29	0.03	-0.17	15	-0.02	-0.22	18
0.23	0.03	3	0.83	0.63	28	0.03	-0.17	14
0.36	0.16	13	1.39	1.19	33	0.87	0.67	30
0.7	0.5	26	0.68	0.48	25	0.3	0.1	9.5
0.71	0.51	27	0.3	0.1	9.5	0.31	0.11	11
0.13	-0.07	7	-0.11	-0.31	22	0.45	0.25	20
0.15	-0.05	5.5	0.31	0.11	12	-0.26	-0.46	24

$s_+ = 424$, so $z = \dfrac{424 - 280.5}{\sqrt{3132.25}} = \dfrac{143.5}{55.9665} = 2.56$. Since $2.56 \geq 1.96$, we reject H_0 at

significance level .05.

9.

r_1	1	1	1	1	1	1	2	2	2	2	2	2
r_2	2	2	3	3	4	4	1	1	3	3	4	4
r_3	3	4	2	4	2	3	3	4	1	4	1	3
r_4	4	3	4	2	3	2	4	3	4	1	3	1
D	0	2	2	6	6	8	2	4	6	12	10	14

r_1	3	3	3	3	3	3	4	4	4	4	4	4
r_2	1	1	2	2	4	4	1	1	2	2	3	3
r_3	2	4	1	4	1	2	2	3	1	3	1	2
r_4	4	2	4	1	2	1	3	2	3	1	2	1
D	6	10	8	14	16	18	12	14	14	18	18	20

When H_0 is true, each of the above 24 rank sequences is equally likely, which yields the distribution of D when H_0 is true as described in the answer section (e.g., P(D = 2) = P(1243 or 1324 or 2134) = 3/24). Then c = 0 yields $\alpha = \frac{1}{24} = .042$ while c = 2 implies $\alpha = \frac{4}{24} = .167$.

Section 15.2

11. With X identified with pine (corresponding to the smaller sample size) and Y with oak, we wish to test $H_0 : \mu_1 - \mu_2 = 0$ vs. $H_a : \mu_1 - \mu_2 \neq 0$. From Table A.14 with m = 6 and n = 8, H_0 is rejected in favor of H_a at level .05 if either $w \geq 61$ or if $w \leq 90 - 61 = 29$ (the actual α is 2(.021) = .042). The X ranks are 3 (for .73), 4 (for .98), 5 (for 1.20), 7 (for 1.33), 8 (for 1.40), and 10 (for 1.52), so w = 37. Since 37 is neither ≥ 61 nor ≤ 29, H_0 cannot be rejected.

13. Here m = n = 10 > 8, so we use the large-sample test statistic from pp. 680-681. $H_0 : \mu_1 - \mu_2 = 0$ will be rejected at level .01 in favor of $H_a : \mu_1 - \mu_2 \neq 0$ if either $z \geq 2.58$ or $z \leq -2.58$. Identifying X with orange juice, the X ranks are 7, 8, 9, 10, 11, 16, 17, 18, 19, and 20, so w = 135. With

$$\frac{m(m+n+1)}{2} = 105 \text{ and } \sqrt{\frac{mn(m+n+1)}{12}} = \sqrt{175} = 13.22, \ z = \frac{135-105}{13.22} = 2.27. \text{ Because}$$

2.27 is neither ≥ 2.58 nor ≤ -2.58, H_0 is not rejected. $p-value \approx 2(1 - \Phi(2.27)) = .0232$.

15. Let μ_1 and μ_2 denote true average cotanine levels in unexposed and exposed infants, respectively. The hypotheses of interest are $H_0 : \mu_1 - \mu_2 = -25$ vs. $H_a : \mu_1 - \mu_2 < -25$. With m = 7, n = 8, H_0 will be rejected at level .05 if $w \leq 7(7+8+1) - 71 = 41$. Before ranking, -25 is subtracted from each x_I (i.e. 25 is added to each), giving 33, 36, 37, 39, 45, 68, and 136. The corresponding ranks in the combined set of 15 observations are 1, 3, 4, 5, 6, 8, and 12, from which w = 1 + 3 + … + 12 = 39. Because $39 \leq 41$, H_0 is rejected. The true average level for exposed infants appears to exceed that for unexposed infants by more than 25 (note that H_0 would not be rejected using level .01).

Section 15.3

17. n = 8, so from Table A.15, a 95% C.I. (actually 94.5%) has the form $\left(\bar{x}_{(36-32+1)}, \bar{x}_{(32)}\right) = \left(\bar{x}_{(5)}, \bar{x}_{(32)}\right)$. It is easily verified that the 5 smallest pairwise averages are $\frac{5.0+5.0}{2} = 5.00$, $\frac{5.0+11.8}{2} = 8.40$,

$\frac{5.0+12.2}{2} = 8.60$, $\frac{5.0+17.0}{2} = 11.00$, and $\frac{5.0+17.3}{2} = 11.15$ (the smallest average not

involving 5.0 is $\bar{x}_{(6)} = \frac{11.8+11.8}{2} = 11.8$), and the 5 largest averages are 30.6, 26.0, 24.7, 23.95, and 23.80, so the confidence interval is (11.15, 23.80).

19. First, we must recognize this as a paired design; the eight <u>differences</u> (Method 1 minus Method 2) are −0.33, −0.41, −0.71, 0.19, −0.52, 0.20, −0.65, and −0.14. With n = 8, Table A.15 gives c = 32, and a 95% CI for μ_D is $(\bar{x}_{(8(8+1)/2-32+1)}, \bar{x}_{(32)}) = (\bar{x}_{(5)}, \bar{x}_{(32)})$.

Of the 36 pairwise averages created from these 8 differences, the 5th smallest is $\bar{x}_{(5)} = -0.585$, and the 5th-largest (aka the 32nd-smallest) is $\bar{x}_{(32)} = 0.025$. Therefore, we are 94.5% confident the true mean difference in extracted creosote between the two solvents, μ_D, lies in the interval (−.585, .025).

21. $m = n = 5$ and from Table A.16, $c = 21$ and the 90% (actually 90.5%) interval is $\left(d_{ij(5)}, d_{ij(21)}\right)$. The five smallest $x_i - y_j$ differences are $-18, -2, 3, 4, 16$ while the five largest differences are 136, 123, 120, 107, 87 (construct a table like Table 15.5), so the desired interval is (16,87).

Section 15.4

23. Below we record in parentheses beside each observation the rank of that observation in the combined sample.

1:	5.8(3)	6.1(5)	6.4(6)	6.5(7)	7.7(10)	$r_{1.} = 31$
2:	7.1(9)	8.8(12)	9.9(14)	10.5(16)	11.2(17)	$r_{2.} = 68$
3:	5.1(1)	5.7(2)	5.9(4)	6.6(8)	8.2(11)	$r_{3.} = 26$
4:	9.5(13)	1.0.3(15)	11.7(18)	12.1(19)	12.4(20)	$r_{4.} = 85$

H_o will be rejected at level .10 if $k \geq \chi^2_{.10,3} = 6.251$. The computed value of k is

$$k = \frac{12}{20(21)}\left[\frac{31^2 + 68^2 + 26^2 + 85^2}{5}\right] - 3(21) = 14.06.$$ Since $14.06 \geq 6.251$, reject H_o.

25. $H_0 : \mu_1 = \mu_2 = \mu_3$ will be rejected at level .05 if $k \geq \chi^2_{.05,2} = 5.992$. The ranks are 1, 3, 4, 5, 6, 7, 8, 9, 12, 14 for the first sample; 11, 13, 15, 16, 17, 18 for the second; 2, 10, 19, 20, 21, 22 for the third; so the rank totals are 69, 90, and 94. $k = \frac{12}{22(23)}\left[\frac{69^2}{10} + \frac{90^2}{6} + \frac{94^2}{5}\right] - 3(23) = 9.23$. Since $9.23 \geq 5.992$, we reject H_o.

27.

	1	2	3	4	5	6	7	8	9	10	r_i	r_i^2
I	1	2	3	3	2	1	1	3	1	2	19	361
H	2	1	1	2	1	2	2	1	2	3	17	289
C	3	3	2	1	3	3	3	2	3	1	24	576
												1226

The computed value of F_r is $\frac{12}{10(3)(4)}(1226) - 3(10)(4) = 2.60$, which is not $\geq \chi^2_{.05,2} = 5.992$, so don't reject H_o.

Supplementary Exercises

29. Friedman's test is appropriate here. At level .05, H_o will be rejected if $f_r \geq \chi^2_{.05,3} = 7.815$. It is easily verified that $r_{1.} = 28$, $r_{2.} = 29$, $r_{3.} = 16$, $r_{4.} = 17$, from which the defining formula gives $f_r = 9.62$ and the computing formula gives $f_r = 9.67$. Because $f_r \geq 7.815$, $H_0 : \alpha_1 = \alpha_2 = \alpha_3 = \alpha_4 = 0$ is rejected, and we conclude that there are effects due to different years.

31. From Table A.16, m = n = 5 implies that c = 22 for a confidence level of 95%, so $mn - c + 1 = 25 - 22 = 1 = 4$. Thus the confidence interval extends from the 4th smallest difference to the 4th largest difference. The 4 smallest differences are –7.1, -6.5, -6.1, -5.9, and the 4 largest are –3.8, -3.7, -3.4, -3.2, so the C.I. is (-5.9, -3.8).

33.

a. With "success" as defined, then Y is a binomial with n = 20. To determine the binomial proportion "p" we realize that since 25 is the hypothesized median, 50% of the distribution should be above 25, thus p = .50. From the Binomial Tables (Table A.1) with n = 20 and p = .50, we see that $\alpha = P(Y \geq 15) = 1 - P(Y \leq 14) = 1 - .979 = .021$.

b. From the same binomial table as in **a**, we find that $P(Y \geq 14) = 1 - P(Y \leq 13) = 1 - .942 = .058$ (as close as we can get to .05), so c = 14. For this data, we would reject H_o at level .058 if $Y \geq 14$. $Y = $ (the number of observations in the sample that exceed 25) = 12, and since 12 < 14, we fail to reject H_o.

35.

Sample:	y	x	y	y	x	x	x	y	y
Observations:	3.7	4.0	4.1	4.3	4.4	4.8	4.9	5.1	5.6
Rank:	1	3	5	7	9	8	6	4	2

The value of W' for this data is $w' = 3 + 6 + 8 + 9 = 26$. At level .05, the critical value for the upper-tailed test is (Table A.14, m = 4, n = 5) c = 27 ($\alpha = .056$). Since 26 is not ≥ 27, H_0 cannot be rejected at level .05.

CHAPTER 16

Section 16.1

1. All ten values of the quality statistic are between the two control limits, so no out-of-control signal is generated.

3. $P(10$ successive points inside the limits$) = P(1^{st}$ inside$)$ x $P(2^{nd}$ inside$)$ x...x $P(10^{th}$ inside$) = (.998)^{10} =$ .9802. $P(25$ successive points inside the limits$) = (.998)^{25} = .9512.$ $(.998)^{52} = .9011,$ but $(.998)^{53} = .8993,$ so for 53 successive points the probability that at least one will fall outside the control limits when the process is in control is $1 - .8993 = .1007 > .10.$

5.

    a.   For the case of 4(a), with $\sigma = .02$, $C_p = \dfrac{USL - LSL}{6\sigma} = \dfrac{3.1 - 2.9}{6(.02)} = 1.67.$ This is indeed a very good

        capability index. In contrast, the case of 4(b) with $\sigma = .05$ has a capability index of $C_p = \dfrac{3.1 - 2.9}{6(.05)} =$

        0.67. This is quite a bit less than 1, the dividing line for "marginal capability."

    b.   For the case of 4(a), with $\mu = 3.04$ and $\sigma = .02$, $\dfrac{USL - \mu}{3\sigma} = \dfrac{3.1 - 3.04}{3(.02)} = 1$ and $\dfrac{\mu - LSL}{3\sigma} = \dfrac{3.04 - 2.9}{3(.02)} =$

        2.33, so $C_{pk} = \min\{1, 2.33\} = 1.$

        For the case of 4(b), with $\mu = 3.00$ and $\sigma = .05$, $\dfrac{USL - \mu}{3\sigma} = \dfrac{3.1 - 3.00}{3(.05)} = .67$ and $\dfrac{\mu - LSL}{3\sigma} = \dfrac{3.00 - 2.9}{3(.05)} =$

        .67, so $C_{pk} = \min\{.67, .67\} = .67.$ Even using this mean-adjusted capability index, process (a) is more "capable" than process (b), though C_{pk} for process (a) is now right at the "marginal capability" threshold.

    c.   In general, $C_{pk} \leq C_p$, and they are equal iff $\mu = \dfrac{LSL + USL}{2}$, i.e. the process mean is the midpoint of the

        spec limits. To demonstrate this, suppose first that $\mu = \dfrac{LSL + USL}{2}$. Then

        $\dfrac{USL - \mu}{3\sigma} = \dfrac{USL - (LSL + USL)/2}{3\sigma} = \dfrac{2USL - (LSL + USL)}{6\sigma} = \dfrac{USL - LSL}{6\sigma} = C_p,$ and similarly

        $\dfrac{\mu - LSL}{3\sigma} = C_p.$ In that case, $C_{pk} = \min\{C_p, C_p\} = C_p.$

        Otherwise, suppose μ is closer to the lower spec limit than to the upper spec limit (but between the

        two), so that $\mu - LSL < USL - \mu.$ In such a case, $C_{pk} = \dfrac{\mu - LSL}{3\sigma}.$ However, in this same case $\mu <$

        $\dfrac{LSL + USL}{2}$, from which $\dfrac{\mu - LSL}{3\sigma} < \dfrac{(LSL + USL)/2 - LSL}{3\sigma} = \dfrac{USL - LSL}{6\sigma} = C_p.$ That is, $C_{pk} < C_p.$

        Analogous arguments for all other possible values of μ also yield $C_{pk} < C_p.$

Section 16.2

7.

a. P(point falls outside the limits when $\mu = \mu_0 + .5\sigma$) $= 1 - P\left(\mu_0 - \dfrac{3\sigma}{\sqrt{n}} < \bar{X} < \mu_0 + \dfrac{3\sigma}{\sqrt{n}} \text{ when } \mu = \mu_0 + .5\sigma\right)$

$= 1 - P\left(-3 - .5\sqrt{n} < Z < 3 - .5\sqrt{n}\right) = 1 - P\left(-4.12 < Z < 1.882\right) = 1 - .9699 = .0301$.

b. $1 - P\left(\mu_0 - \dfrac{3\sigma}{\sqrt{n}} < \bar{X} < \mu_0 + \dfrac{3\sigma}{\sqrt{n}} \text{ when } \mu = \mu_0 - \sigma\right) = 1 - P\left(-3 + \sqrt{n} < Z < 3 + \sqrt{n}\right)$

$= 1 - P\left(-.76 < Z < 5.24\right) = .2236$

c. $1 - P\left(-3 - 2\sqrt{n} < Z < 3 - 2\sqrt{n}\right) = 1 - P\left(-7.47 < Z < -1.47\right) = .9292$

9. $\bar{\bar{x}} = 12.95$ and $\bar{s} = .526$, so with $a_5 = .940$, the control limits are

$12.95 \pm 3\dfrac{.526}{.940\sqrt{5}} = 12.95 \pm .75 = 12.20, 13.70$. Again, every point $(\bar{x})$ is between these limits, so there is no evidence of an out-of-control process.

11. $\bar{\bar{x}} = \dfrac{2317.07}{24} = 96.54$, $\bar{s} = 1.264$, and $a_6 = .952$, giving the control limits

$96.54 \pm 3\dfrac{1.264}{.952\sqrt{6}} = 96.54 \pm 1.63 = 94.91, 98.17$. The value of $\bar{x}$ on the 22nd day lies above the UCL, so the process appears to be out of control at that time.

13.

a. $P\left(\mu_0 - \dfrac{2.81\sigma}{\sqrt{n}} < \bar{X} < \mu_0 + \dfrac{2.81\sigma}{\sqrt{n}} \text{ when } \mu = \mu_0\right) = P\left(-2.81 < Z < 2.81\right) = .995$, so the

probability that a point falls outside the limits is .005 and $ARL = \dfrac{1}{.005} = 200$.

b. p = P(a point is outside the limits) $= 1 - P\left(\mu_0 - \dfrac{2.81\sigma}{\sqrt{n}} < \bar{X} < \mu_0 + \dfrac{2.81\sigma}{\sqrt{n}} \text{ when } \mu = \mu_0 + \sigma\right)$

$= 1 - P\left(-2.81 - \sqrt{n} < Z < 2.81 - \sqrt{n}\right) = 1 - P\left(-4.81 < Z < .81\right) = 1 - .791 = .209$. Thus

$ARL = \dfrac{1}{.209} = 4.78$

c. $1 - .9974 = .0026$ so $ARL = \dfrac{1}{.0026} = 385$ for an in-control process, and when $\mu = \mu_0 + \sigma$, the

probability of an out-of-control point is $1 - P(-3 - 2 < Z < 1) = 1 - P(Z < 1) = .1587$, so

$ARL = \dfrac{1}{.1587} = 6.30$.

15. $\bar{\bar{x}} = 12.95$, IQR = .4273, $k_5 = .990$. The control limits are $12.95 \pm 3 \dfrac{.4273}{.990\sqrt{5}} = 12.37, 13.53$.

Section 16.3

17.

a. $\bar{r} = \dfrac{85.2}{30} = 2.84$, $b_4 = 2.058$, and $c_4 = .880$. Since n = 4, LCL = 0 and UCL

$$= 2.84 + \dfrac{3(.880)(2.84)}{2.058} = 2.84 + 3.64 = 6.48.$$

b. $\bar{r} = 3.54$, $b_8 = 2.844$, and $c_8 = .820$, and the control limits are

$$= 3.54 \pm \dfrac{3(.820)(3.54)}{2.844} = 3.54 \pm 3.06 = .48, 6.60.$$

19. $\bar{s} = 1.2642$, $a_6 = .952$, and the control limits are

$$1.2642 \pm \dfrac{3(1.2642)\sqrt{1-(.952)^2}}{.952} = 1.2642 \pm 1.2194 = .045, 2.484.$$ The smallest s_i is $s_{20} = .75$, and the largest is $s_{12} = 1.65$, so every value is between .045 and 2.434. The process appears to be in control with respect to variability.

Section 16.4

21. $\bar{p} = \Sigma \dfrac{\hat{p}_i}{k}$ where $\Sigma\hat{p}_i = \dfrac{x_1}{n} + ... + \dfrac{x_k}{n} = \dfrac{x_1 + ... + x_k}{n} = \dfrac{578}{100} = 5.78$. Thus $\bar{p} = \dfrac{5.78}{25} = .231$.

a. The control limits are $.231 \pm 3\sqrt{\dfrac{(.231)(.769)}{100}} = .231 \pm .126 = .105, .357$.

b. $\dfrac{13}{100} = .130$, which is between the limits, but $\dfrac{39}{100} = .390$, which exceeds the upper control limit and therefore generates an out-of-control signal.

23. LCL > 0 when $\bar{p} > 3\sqrt{\dfrac{\bar{p}(1-\bar{p})}{n}}$, i.e. (after squaring both sides) $50\bar{p}^2 > 9\bar{p}(1-\bar{p})$, i.e.

$50\bar{p} > 3(1-\bar{p})$, i.e. $53\bar{p} > 3 \Rightarrow \bar{p} = \dfrac{3}{53} = .0566$.

25. $\Sigma x_i = 102$, $\bar{x} = 4.08$, and $\bar{x} \pm 3\sqrt{\bar{x}} = 4.08 \pm 6.06 \approx (-2.0, 10.1)$. Thus LCL = 0 and UCL = 10.1. Because no x_i exceeds 10.1, the process is judged to be in control.

27. With $u_i = \dfrac{x_i}{g_i}$, the u_i's are 3.75, 3.33, 3.75, 2.50, 5.00, 5.00, 12.50, 12.00, 6.67, 3.33, 1.67, 3.75, 6.25,

4.00, 6.00, 12.00, 3.75, 5.00, 8.33, and 1.67 for I = 1, ..., 20, giving $\bar{u} = 5.5125$. For $g_i = .6$,

$$\bar{u} \pm 3\sqrt{\frac{\bar{u}}{g_i}} = 5.5125 \pm 9.0933, \text{LCL} = 0, \text{UCL} = 14.6. \text{ For } g_i = .8, \bar{u} \pm 3\sqrt{\frac{\bar{u}}{g_i}} = 5.5125 \pm 7.857,$$

LCL = 0, UCL = 13.4. For $g_i = 1.0$, $\bar{u} \pm 3\sqrt{\dfrac{\bar{u}}{g_i}} = 5.5125 \pm 7.0436$, LCL = 0, UCL = 12.6. Several

u_i's are close to the corresponding UCL's but none exceed them, so the process is judged to be in control.

Section 16.5

29. $\mu_0 = 16$, $k = \dfrac{\Delta}{2} = 0.05$, $h = .20$, $d_i = \max\left(0, d_{i-1} + (\bar{x}_i - 16.05)\right)$, $e_i = \max\left(0, e_{i-1} + (\bar{x}_i - 15.95)\right)$.

i	$\bar{x}_i - 16.05$	d_i	$\bar{x}_i - 15.95$	e_i
1	-0.058	0	0.024	0
2	0.001	0.001	0.101	0
3	0.016	0.017	0.116	0
4	-0.138	0	-0.038	0.038
5	-0.020	0	0.080	0
6	0.010	0.010	0.110	0
7	-0.068	0	0.032	0
8	-0.151	0	-0.054	0.054
9	-0.012	0	0.088	0
10	0.024	0.024	0.124	0
11	-0.021	0.003	0.079	0
12	-0.115	0	-0.015	0.015
13	-0.018	0	0.082	0
14	-0.090	0	0.010	0
15	0.005	0.005	0.105	0

For no time r is it the case that $d_r > .20$ or that $e_r > .20$, so no out-of-control signals are generated.

31. Connecting 600 on the in-control ARL scale to 4 on the out-of-control scale and extending to the k' scale

gives k' = .87. Thus $k' = \dfrac{\Delta/2}{\sigma/\sqrt{n}} = \dfrac{.002}{.005/\sqrt{n}}$ from which $\sqrt{n} = 2.175 \Rightarrow n = 4.73 = s$. Then

connecting .87 on the k' scale to 600 on the out-of-control ARL scale and extending to h' gives h' = 2.8, so

$$h = \left(\frac{\sigma}{\sqrt{n}}\right)(2.8) = \left(\frac{.005}{\sqrt{5}}\right)(2.8) = .00626.$$

Section 16.6

33. For the binomial calculation, n = 50 and we wish

$$P(X \le 2) = \binom{50}{0} p^0 (1-p)^{50} + \binom{50}{1} p^1 (1-p)^{49} + \binom{50}{2} p^2 (1-p)^{48}$$

$$= (1-p)^{50} + 50p(1-p)^{49} + 1225 p^2 (1-p)^{48} \text{ when } p = .01, .02, \ldots, .10. \text{ For the hypergeometric}$$

calculation $P(X \le 2) = \dfrac{\binom{M}{0}\binom{500-M}{50}}{\binom{500}{50}} + \dfrac{\binom{M}{1}\binom{500-M}{49}}{\binom{500}{50}} + \dfrac{\binom{M}{2}\binom{500-M}{48}}{\binom{500}{50}}$, to be calculated

for M = 5, 10, 15, ..., 50. The resulting probabilities appear in the answer section in the text.

35. $P(X \le 2) = \binom{100}{0} p^0 (1-p)^{100} + \binom{100}{1} p^1 (1-p)^{99} + \binom{100}{2} p^2 (1-p)^{98}$

p	.01	.02	.03	.04	.05	.06	.07	.08	.09	.10
$P(X \le 2)$	.9206	.6767	.4198	.2321	.1183	.0566	.0258	.0113	.0048	.0019

For values of p quite close to 0, the probability of lot acceptance using this plan is larger than that for the previous plan, whereas for larger p this plan is less likely to result in an "accept the lot" decision (the dividing point between "close to zero" and "larger p" is someplace between .01 and .02). In this sense, the current plan is better.

37. P(accepting the lot) = P(X₁ = 0 or 1) + P(X₁ = 2, X₂ = 0, 1, 2, or 3) + P(X₁ = 3, X₂ = 0, 1, or 2) = P(X₁ = 0 or 1) + P(X₁ = 2)P(X₂ = 0, 1, 2, or 3) + P(X₁ = 3)P(X₂ = 0, 1, or 2).

$p = .01:\ = .9106 + (.0756)(.9984) + (.0122)(.9862) = .9981$
$p = .05:\ = .2794 + (.2611)(.7604) + (.2199)(.5405) = .5968$
$p = .10:\ = .0338 + (.0779)(.2503) + (.1386)(.1117) = .0688$

39.

a. $AOQ = pP(A) = p[(1-p)^{50} + 50p(1-p)^{49} + 1225 p^2 (1-p)^{48}]$

p	.01	.02	.03	.04	.05	.06	.07	.08	.09	.10
AOQ	.010	.018	.024	.027	.027	.025	.022	.018	.014	.011

b. p = .0447, AOQL = .0447P(A) = .0274

c. ATI = 50P(A) + 2000(1 − P(A))

p	.01	.02	.03	.04	.05	.06	.07	.08	.09	.10
ATI	77.3	202.1	418.6	679.9	945.1	1188.8	1393.6	1559.3	1686.1	1781.6

Supplementary Exercises

41. $n = 6, k = 26, \Sigma \overline{x}_i = 10{,}980, \overline{\overline{x}} = 422.31, \Sigma s_i = 402, \overline{s} = 15.4615, \Sigma r_i = 1074, \overline{r} = 41.3077$

S chart: $15.4615 \pm \dfrac{3(15.4615)\sqrt{1 - (.952)^2}}{.952} = 15.4615 \pm 14.9141 \approx .55, 30.37$

R chart: $41.31 \pm \dfrac{3(.848)(41.31)}{2.536} = 41.31 \pm 41.44$, so LCL = 0, UCL = 82.75

$\overline{X}$ chart based on $\overline{s}$: $422.31 \pm \dfrac{3(15.4615)}{.952\sqrt{6}} = 402.42, 442.20$

$\overline{X}$ chart based on $\overline{r}$: $422.31 \pm \dfrac{3(41.3077)}{2.536\sqrt{6}} = 402.36, 442.26$

43.

i	$\overline{x}_i$	s_i	r_i
1	50.83	1.172	2.2
2	50.10	.854	1.7
3	50.30	1.136	2.1
4	50.23	1.097	2.1
5	50.33	.666	1.3
6	51.20	.854	1.7
7	50.17	.416	.8
8	50.70	.964	1.8
9	49.93	1.159	2.1
10	49.97	.473	.9
11	50.13	.698	.9
12	49.33	.833	1.6
13	50.23	.839	1.5
14	50.33	.404	.8
15	49.30	.265	.5
16	49.90	.854	1.7
17	50.40	.781	1.4
18	49.37	.902	1.8
19	49.87	.643	1.2
20	50.00	.794	1.5
21	50.80	2.931	5.6
22	50.43	.971	1.9

$\Sigma s_i = 19.706, \overline{s} = .8957, \Sigma \overline{x}_i = 1103.85, \overline{\overline{x}} = 50.175, a_3 = .886$, from which an s chart has

LCL = 0 and UCL = $.8957 + \dfrac{3(.8957)\sqrt{1 - (.886)^2}}{.886} = 2.3020$, and $s_{21} = 2.931 > UCL$. Since an

assignable cause is assumed to have been identified we eliminate the 21st group. Then $\Sigma s_i = 16.775$,

$\bar{s} = .7998$, $\bar{\bar{x}} = 50.145$. The resulting UCL for an s chart is 2.0529, and $s_i < 2.0529$ for every remaining i. The $\bar{x}$ chart based on $\bar{s}$ has limits $50.145 \pm \dfrac{3(.7988)}{.886\sqrt{3}} = 48.58, 51.71$. All $\bar{x}_i$ values are between these limits.

45. $\Sigma n_i = 4(16) + (3)(4) = 76$, $\Sigma n_i \bar{x}_i = 32,729.4$, $\bar{\bar{x}} = 430.65$,

$$s^2 = \frac{\Sigma(n_i - 1)s_i^2}{\Sigma(n_i - 1)} = \frac{27,380.16 - 5661.4}{76 - 20} = 590.0279, \text{ so } s = 24.2905. \text{ For variation: when } n = 3,$$

$$UCL = 24.2905 + \frac{3(24.2905)\sqrt{1 - (.886)^2}}{.886} = 24.29 + 38.14 = 62.43, \text{ when } n = 4,$$

$$UCL = 24.2905 + \frac{3(24.2905)\sqrt{1 - (.921)^2}}{.921} = 24.29 + 30.82 = 55.11. \text{ For location: when } n = 3,$$

$430.65 \pm 47.49 = 383.16, 478.14$, and when n = 4, $430.65 \pm 39.56 = 391.09, 470.21$.